Jerome A. Berson

Chemical Creativity

WILEY-VCH

Jerome A. Berson

Chemical Creativity

Ideas from the Work of Woodward, Hückel, Meerwein, and Others

Weinheim · New York · Chichester · Brisbane · Singapore · Toronto

Prof. J. A. Berson
Department of Chemistry
Yale University
225 Prospect Street, P.O. Box 208107
New Haven, CT 06520-8107
USA

1st edition, 1999

The cover illustration shows the American chemist and Nobel Prize-winner Robert B. Woodward (photograph from the Harvard University Archives). Reproduced with the kind permission of Ms. Crystal Woodward.

Library of Congress Card No.: applied for.

A catalogue record for this book is available from the British Library.

Deutsche Bibliothek Cataloguing-in-Publication Data:
Berson, Jerome A.:
Chemical creativity : ideas from the work of Woodward, Hückel, Meerwein, and others / Jerome A. Berson. - Weinheim ; New York ; Chichester ; Brisbane ; Singapore ; Toronto : Wiley-VCH, 1999
ISBN 3-527-29754-5

Printed on acid-free and chlorine-free paper.

Cover Design: Gunther Schulz, D-67136 Fußgönheim
Composition: Datascan GmbH, D-67346 Speyer
Printing: Strauss Offsetdruck, D-69509 Mörlenbach
Bookbinding: Wilhelm Osswald & Co., D-67433 Neustadt/Weinstraße

Printed in the Federal Republic of Germany.

To Bella

Preface

A Note to the Reader. I hope that this book will be read by scientists from a broad range of disciplines and by others interested in the genesis of ideas. Although some of the subject matter is technical, I do not think the reader is required to be expert in organic chemistry or quantum mechanics to engage the main themes. Throughout the text, I have included summaries of the significance of the technical passages just completed, so that the non-specialist reader can follow the broad outlines of the arguments. If I may judge from the reaction of reviewers of the manuscript, I believe that those readers who are ready to plunge into the more specific aspects will find the technical portions of the book to be stimulating, provocative and occasionally perhaps controversial.

I refer frequently to two major German chemical journals, both of which have undergone significant changes of name over the decades of their separate existence. Thus, the journal now entitled *European Journal of Organic Chemistry* previously was known as *Liebigs Annalen der Chemie*, and earlier, as *Justus Liebigs Annalen der Chemie*. The present-day *European Journal of Inorganic Chemistry* was previously known as *Chemische Berichte* and before that as *Berichte der Deutschen Chemischen Gesellschaft*. For simplicity, I refer to these in the lists of references as *Ann.* and *Ber.*, respectively.

Acknowledgments. While writing this book, I have relied upon numerous colleagues who have given generously of their time, expertise, and advice, especially in areas outside my personal experience. In this respect, the book is really a collaborative project. I have acknowledged their contributions at the ends of the appropriate chapters. The interaction with these and other consultants has taught me a great deal, but any remaining misconceptions, errors, or other flaws are entirely my own responsibility.

There are several persons whose influence on the final shape of the book has been especially noteworthy. On behalf of the publisher, the complete manuscript has been read by the editor, Dr. Peter Gölitz, and also by Professors Henning Hopf, Wolfgang Lüttke, and Lionel Salem. Each of these readers made extensive suggestions and provided many useful documents and references, as well as astute written comments, most of which have been incorporated. Professor Kurt Mislow went far beyond the call of duty to provide a penetrating examination of the complete manuscript, many insightful suggestions for improvements, and hours of animated discussion. My wife, Bella Berson, also read the manuscript while it was in process and made me think as deeply as I could about what I was trying to say. She also contributed two other important things: her advice, based upon deep experience, about how to use library rsources efficiently, and endless patience. Professor Roald Hoffmann not only gave expert advice and personal recollections about the subject matter of some of the chapters but also encouraged my early ambitions to write on the history of chemistry, despite my lack of formal background in historical studies.

Chapter 3 contains material on the life of Erich Hückel which is not available in the chemical literature. I am especially grateful to Professor W. Walcher of the University of Marburg and to Professor I. Auerbach, chief archival consultant of the Hessischen Staatsarchiv in that city, for the valuable information they provided about the circumstances of Hückel's appointment to the faculty there. I am deeply indebted to Professor Horst Tietz of the University of Hannover for his account of the origin and course of his relationship with Hückel, for a description of the experiences which the Tietz family endured during the period 1933–1945, and for permission to disclose this material here.

For permission to reproduce text or graphic materials, I thank Oxford University Press, WILEY-VCH Verlag GmbH, The Royal Society of Chemistry, The Royal Society, and Elsevier Science Ltd. Acknowledgments of specific materials are given at places where they appear in the book.

New Haven, Connecticut

August, 1998

Table of Contents

Chapter 1

Introduction

»... I consider the teaching and study of the historical development of science as indispensable ... Our textbooks fail in this respect.«

Richard Willstätter[1]

»... Science is not an abstract thing, but rather, as a product of human labor, is tightly bound in its development to the particularity and fate of the individuals who dedicate themselves to it.«

Emil Fischer[1]

1.1 Objectives

The history of chemistry often is written from an elevated perspective. The historian seeks to identify prominent innovators and their schools of co-workers, to trace intersecting lines of influence, to analyze large-scale trends in the development of the science, and to demonstrate interactions with biology, physics, technology, and the world at large. There is little doubt that many historians of science will continue to employ the breadth of view inherent in these methods, which have excelled in locating chemistry in the intellectual landscape and in making available some of the major results of chemical research to chemists as well as to non-specialist audiences.

The present essays have a different set of objectives. They are informed by my unproven but nevertheless strongly held conviction that an awareness of the details of how chemical ideas develop – a knowledge, so to say, of how we came to know what we know – can make us better, more thoughtful, more creative scientists.

The book is not a textbook or a research monograph. Thus, the objective is not to bring the reader to the frontier of today's research. Instead, these essays seek out the *origins* of ideas. They analyze the historical record of specific research problems for the original motivations of the investigators, which often were distinct from the ultimate significance recognized only later. They examine the bright (if sometimes oversimplified or even false) notions that sparked the investigators' imagination, they follow details of the experimental search for understanding, tracing not only what was done (sometimes incompletely) but how it was interpreted (sometimes too optimistically). They point out experiments that should have been done and thoughts that should have been thought (but sometimes were not). The point of view is not above but rather at ground level. In this way, I hope to engage the reader almost as a participant in the development of a significant piece of research. If I am successful, the reader will experience vicariously the day-to-day activity in the mind of the chemist: the long hours of tedious labor, the frustration of an uninformative result, the brain-racking laboratory conferences in search of solutions, and the rare excitement that comes with the first glimmer of insight. The objective is not to display what (we think) we know now, nor by hindsight-aided iconoclasm, to show how much wiser we are than our distinguished predecessors, but rather to reveal how chemists think. I am convinced that inspection from this angle gives us insights that apply to any field of intellectual endeavor.

1.2 What is Special About This Book?

Of course, the detailed case-study method is not unprecedented. There is a large literature of articles and books on case-studies in science. Among the many that could be cited are individual papers about specific discoveries in chemistry[2] as well as collected accounts from particle physics and other disciplines[3–6] In fact, in some respects, the present work has been stimulated by efforts initiated at Harvard by James B. Conant and first outlined publicly in his Terry Lectures at Yale in 1946.[7] Conant wrote

>> *This book is concerned primarily with a simple yet difficult pedagogic problem. I propose to examine the question of how we can in our colleges give a better understanding of science to those of our graduates who are to be lawyers, writers, teachers, politicians, public servants, and businessmen. To the extent that my answer to this query has novelty, I shall be forced to illustrate by example. Since it is my contention that science can best be understood by laymen through close study of a few relatively simple case histories, I have no choice but to present some fragments of scientific history … The objectives (of the new approach) would be to give a greater degree of understanding of science by the close study of a relatively few historical examples of the development of science.* << [7a]

An effort to incorporate these ideas began at Harvard 1947 in an experimental course designed for those freshmen and sophomores who "will neither concentrate in a natural science nor use physics or chemistry in connection with a professional career."[7b]

So why is another case-study book needed? One reason is that I aim at a different readership than Conant, not the indifferent (or perhaps actually hostile) freshmen and sophomores he was trying to reach, but rather advanced students and working professionals in chemistry and allied fields. It is true that my essays share with the Harvard project the idea of developing insight about the strategy and tactics of chemical research by the examination of specific cases. A crucial difference, however, is that those who are most likely to read this book already have a great deal of sophistication and detailed knowledge of the subject. The objective thus is not to generate in them an appreciation for science; that they already have in abundance, having dedicated their lives to it. To benefit from this effort, they need to perceive, or at least to be open to the perception, that an understanding of the history of the subject can provide lessons that will be applicable to their own contemporary work.

Detailed accounts of chemical discoveries often can be found in the autobiographies of chemists. Some of these give illuminating glances into the minds of working scientists.[8] These reminiscences are easier to produce than the hard-won exposition of the independent writer, who must dig out the facts from the literature, from archives of documentation or correspondence, and from direct consultation of other individuals acquainted with the work under study and its proponents. With due deference to the autobiographers, one needs to be aware that not all of our colleagues, even (or, some would say, especially!) the most distinguished, can resist a natural inclination to cosmeticize the historical record for self-serving purposes. In the words of Duff Cooper[9] as quoted by Willstätter:[10]

»*Any reader is at liberty to believe as much or as little of contemporary accounts as he desires, and indeed half the fascination of studying the memoirs of the past is the endeavor, by making allowance for the prejudices and predilections of the writer, to sift truth from falsehood.*«

Rearranging the facts of the past to suit one's own purposes seems to be a common infection to which, objective introspection tells me, I am not immune. Therefore, although I have chosen for this book examples which are certainly related in a broad sense to my personal research interests over the years, I have concentrated on the efforts of others. I hope that in doing so, I have been able to avoid conflicts of interest and to allot criticism and praise where they are deserved.

Another special characteristic of this book is that its choice of topics is not motivated primarily by an attempt to cover particular periods or particular subjects in the history of chemistry. Its range of subject matter treats some of the most sublime ideas of the last 100 years, but it also includes more mundane material. Thus, rather than the display of brilliant examples to show how the whole field advanced, our method consists in the analysis of how individual investigators chose problems and tried to solve them – not how chemical research *should be* done, but rather how it *was* done. In the process, we shall find that whatever claims to "objectivity" scientists may make, the personal histories and circumstances of the protagonists can be decisive in the development of a field.

Similarly, the researchers we shall meet include some of the greatest chemists but also some less than great. From members of both categories we encounter clear thinking and inspired experimentation as well as blunders, oversights, and misconceptions, including some by demigods of the science. The purpose here is not to entertain with tales of the shortcomings of our heroes, or to make the trivial point that we are all human. Nor is it to give us the satisfaction of realizing that those errors could not be made today because, after all, we now understand the issue in the light of modern knowledge. The point is that as we carry out our own researches today, we are also surrounded by our own set of different, contemporary, and unrecognized false assumptions. The lesson will be that by imagining ourselves in the intellectual surroundings of former times, we may come to understand the reasons why the paths to understanding are invariably forked, crooked, and rough, and we may become more aware of the incompleteness of our own knowledge.

We must remember, however, that the experiences of history, no matter how thoroughly we incorporate them into our consciousness, do not offer useful *rules* for procedure in the present. Feyerabend,[11] points out this problem with an apt quotation from Mach:

»*It is often said that research cannot be taught. That is quite correct, in a certain sense. The schemata of* formal *logic and of* inductive *logic are of little use for the intellectual situations are never exactly the same. But the examples of great scientists are very suggestive.*«[12]

In a passage that comes close to capturing the essence of this book, Feyerabend continues:

»*They are not suggestive because we can abstract rules from them and subject future research to their jurisdiction; they are suggestive because they make the mind nimble and capable of inventing entirely new research traditions.*«[11]

My own experience tells me that these these concepts are correct. It is for this reason that one must be skeptical of the admonitory mode sometimes used to teach young researchers the strategic and tactical skills they will need to flourish as creative scientists:

1. Don't forget to do the right control experiments.
2. Keep in mind the possibility that your interpretation of your results may not be unique.
3. Be careful not to let your research be driven by the demands of a technique rather than by scientific importance.
4. Stay alert to the possibility that the puzzling observation you have just made may be the answer to a question that has not yet been asked.
5. See the big picture, but sweat the details too.
6. Until it is recognized, a blunder bears an uncanny resemblance to a discovery.
7. The popularity of a problem is not necessarily a measure of its importance. Create *your* problem.

And so forth. Such rules can hardly be disputed, but for the reasons just given, they offer no automatic guidelines applicable to a new situation. I would contend that even if you are a teacher whose own research accomplishments merit respect, admiration, and emulation, and even if your beginning students post such a list on their bulletin boards, or memorize them, the crucial moment when one of your maxims becomes relevant to an event in the laboratory is likely to escape their notice.

For that reason, among others, the Ph. D. research project, in which the student, under the mentor's guidance, is forced by the nature of the research itself to confront these issues frequently and thereby to learn by experience, has become an important component of graduate education. Of course, the range of subject matter, and hence the breadth of experience in most individual Ph. D. projects, is necessarily narrow and limited. Therefore, one might hope to supplement the Ph. D. research project with examples of crucial moments taken from the vast collective experience embodied in the chemical literature. But what Conant pointed out fifty years ago is still true: these stories are not told in standard courses. We teachers are hard-pressed to organize, convey, and explain an ever-increasing mass of information for students whose progress toward professional status simply demands that they master the factual and conceptual basis of their subject. Although I, and no doubt others, have tried to incorporate occasional historical analyses (beyond mere anecdotes), we find that there is not enough time to do so regularly in standard lecture-format courses. This is where I hope this book can be helpful.

It is true that this kind of analysis often can come into play informally in the leisure of a late-night literature seminar or in the innumerable conferences between professor and student in the laboratory when the work of others becomes relevant to their common objective. However, the insights needed to extract the lessons of experience we seek to impart may not emerge spontaneously even from the raw literature itself. The typical scientific paper has its own agenda, chief among which is the aim to convince the reader of its validity in the narrow sense of the specific problem it addresses. This urgency, in my opinion, bears a large portion of the blame for the often criticized style characteristic of most scientific papers: bland, stilted, just-the-facts, dead-pan. Editors of scientific journals, most of whom after all are themselves working scientists, have come to discourage authors from taking the time *or journal space* to set the research in the context we are pursuing here. The stories in this book should serve as examples of how one can trace, from a group of papers, supplemented by other sources, the emergence of an important generalization.

1.3 The Nature of Science and the History of Science

Before beginning these accounts, I must digress to treat a question that is sure to be asked: should this book be considered part of the intellectual discipline of history of science? Or are these stories supposed to serve only a pedagogical purpose in educating research chemists? This in turn raises the question of what we mean by "history of science."

At the beginning of this introduction, I outlined some of the major themes that have been pursued by legitimate workers in that field. There is no doubt that in order to do justice to the interactions of science with the world, a historian of science must have command of many branches of the humanities and social sciences. Indeed, there are those in academia who, if I understand them, seem to think that *all* science is socially and historically determined. I would not deny the strong influence of social, political, and economic forces on the development of science. For example, one sees that this must inevitably be the case in modern times, when the cost of doing science is so large that it cannot be born by the individual scientist, but must be approved or at least tolerated by society at large. In blunter words, if we would stay in science as professionals, we cannot ignore the value systems of those who support our work.

More generally, the dedicated scientist brings to his or her work a mind, a personality, and a set of beliefs shaped by the events of a lifetime. To the extent that we are the products of our own histories, so too is our science the product of those histories.

But there is more to it than that. There are those who believe that scientists are totally misguided, if not actually disingenuous, in claiming to be the arbiters of what is "true," when they should admit that scientific beliefs are merely "texts" with no more claim to "truth" than other texts derived by non-scientific means. Arguing this issue here is not my purpose, but let me state that few working scientists of my acquaintance spend much of their time worrying about the philosophical nature of truth. For our needs, it suffices to know that a particular formulation of experimental and theoretical findings is explicative and predictive. Although probably few of us, when sober, would subscribe to the boozy nihilism of Yeats's verse[13]

> »*Wine, comes in at the mouth,*
> *And love comes in at the eye.*
> *That's all we shall know for truth*
> *Before we grow old and die* ...«,

nevertheless we see a strong utilitarian component to the acceptability of a scientific concept.

One can illustrate this with an example from the application of science to human purpose. Thus, equipped with Newton's laws of motion, an accurate knowledge of the earth's surface, and the rate of fuel consumption derived from the principles of physical chemistry, our airplane pilot can take a loaded jumbo jet aloft from San Francisco and hit that tiny spot called Honolulu in the middle of the Pacific Ocean unerringly and safely every time. In this process, science has fulfilled its predictive obligation, whether or not the laws are "true" by some philosophical definition. In fact, we now believe that Newton's laws of motion are merely a limiting case of the Einsteinian relativity relationships, and so, in some sense, not "true" at all. But any scientist knows also that the relativistic deviation from Newtonian behavior in our flight to Honolulu will be too small to detect. That we can conduct our everyday terrestrial lifestyle as if Newton's laws were in fact true illustrates the useful concept of limiting cases in science, a way of thinking which distinguishes between "true" and "true enough."

My attitude, and I dare say that of most other chemists, toward the issue of "truth,"even in pure science itself, is informed by the same pragmatism. For example, Erich Hückel's theory (see Chapter 3) of π-electrons in cyclic conjugated molecules was not really "true," as he certainly recognized. It was a gross approximation to the full solution of the Schrödinger equation. Yet it was extremely useful in explaining the phenomenon of aromaticity and in predicting the existence of new forms of matter, many of which were subsequently synthesized.

Much of the foregoing digression was to show that in practice, the efforts of most working scientists are not aimed at finding absolute "truth." The program of science, rather, is to find that bit of conceptualization which allows us to take the next step forward, to explain something we didn't understand before, to predict that if we mix A with B we'll get C, not D. That the concept we use may later have to be modified or abandoned in favor of a more comprehensive or completely different theory, that is, that we didn't really "understand" at all, is not a reason to avoid using the imperfect theory now if we have no other way to proceed. In other words, all of theory is provisional in some sense. Its major purpose is to bring order, clarity, and predictability to a small corner of the world. We need be neither saddened nor shamed by the realization that achievement of this goal may be – in fact, is likely to be – evanescent. A lot of science consists of permanent experimental facts established in tests of temporary theories.

The recorded literature of these myriad efforts allows us to trace the issues that have driven the dynamics of the scientific community. The case studies in this book seek to illustrate the nature of this complicated set of human interactions. They show how science is actually done. It is in this sense that I believe that they are a legitimate part of history of science.

1.4 This Book: An Experiment

This book therefore may be considered an experiment to test the didactic influence of that branch of history of science on the discipline of chemistry. The standard accepted concepts of chemistry, learned with little or no historical development in the usual formal systematic presentation of our conventional courses, rarely will suffice to sustain a career of significant discovery for the chemist of today. I hope that the following stories will provide some of the "suggestive" examples that Mach and Feyerabend had in mind. If I succeeed, they will help to prepare chemists to confront the unfamiliar, to recognize the difference between a breakthrough and a stupid mistake, to realize that there is much that we still don't know, to look upon surprises in science as events to be welcomed, not feared, and to understand that many of the fundamental ideas of our discipline arose from dubious reasoning, only to be purified in later refinements to their current shining state.

Another way of expressing such motivations appears in an article by Gerald Holton:[14]

» *The progress of science is threatened today not only by loss of financial support and of good people – which is bad enough; not only by diversion of too much of its energy to applied or engineering work that may not yet be bolstered by enough basic knowledge; and not only by the confusion and disenchantment of the wider public – and that, too, demands our concern, because some of it is surely due to lack of proper attention on the part of scientists. No, what seems to me to be the most sensitive, the most fragile part of the total intellectual ecology of science is the understanding, on the part of scientists themselves, of the nature of the scientific enterprise, and in particular the hardly begun study of the nature of scientific discovery. In this pursuit, our own day-by-day experience as scientists will help us if we set it into the historic framework provided by those who went before us.* «

1.5 Acknowledgments

I thank W. Lüttke and K. Mislow for references and helpful suggestions.

1.6 References

(1) As quoted by Huisgen, R. *Angew. Chem.* **1986**, *98*,,297. I think W. Lüttke for calling this quotation to my attention.
(2) Examples include (a) the hexaphenylethane riddle: McBride, J.M. *Tetrahedron*, **1974**, 30, 2009. (b) Salem's theory of organic photochemistry: Turro, N.J. *J. Mol. Struct. (Theochem)*, **1998**, *424*, 77. (c) Pasteur's discovery of molecular asymmetry: Bernal, J.D. *Science and Industry in the Nineteenth Century*, Routledge and Kegan Paul Ltd., London, 1953. (d) the origins of Robinson's synthesis of tropinone: Birch, A.J. J. *Proc. R. Soc. New South Wales,* **1976**, 109, 151. (e) Baeyer's synthesis of indigo: Willstätter, R. *Aus Meinem Leben*, Verlag Chemie, Weinheim, 1949, p. 124 ff.(f) Harré, R. *Great Scientific Experiments: Twenty Experiments That Changed Our View of the World*, Oxford University Press, New York, 1983.
(3) Holton, G. *The Scientific Imagination, Case Studies*, Cambridge University Press, New York, NY, 1978.
(4) Gooding, D.; Pinch, T.; Schaffer, S. *The Uses of Experiment: Studies in the Natural Sciences*, Cambridge University Press, Cambridge, 1989.
(5) Engelhardt, Jr., T.; Caplan, A.L., eds. *Scientific Controversies Case Studies in Resolution and Closure of Dispute in Science and Technology*, Cambridge University Press, 1983.
(6) Nickles, T. *Scientific Discovery, Case Studies*, Kluwer, Boston, MA, 1980.
(7) (a) Conant, J.B. *On Understanding Science. An Historical Approach* (the Terry Lectures at Yale University), Yale University Press, New Haven, 1946, p. 1. (b) *ibid.* p. 15. (c) *ibid.* p. 11. Conant, J.B. *The Growth of the Experimental Sciences. An Experiment in General Education. Progress Report on the Use of the Case Method in Teaching the Principles of the Tactics and Strategy of Science*, Harvard University Press, Cambridge, MA, 1949. (d) Conant, J.B. in *Harvard Case Histories in Experimental Science*, Conant, J.B., ed.; Nash, L.K., assoc. ed. Harvard University Press, Cambridge, MA, 1957, Vol. I, p. ix.
(8) An exhilarating example is Nozoe's account of the discovery of the troponoids in Nozoe, T. *Seventy Years in Organic Chemistry*, American Chemical Society, (Seeman, J.I., ed.), Washington, D.C., 1991.
(9) Cooper, Duff, Viscount Norwich, *Talleyrand*, Harper, New York, 1932.
(10) Willstätter, R. *Aus Meinen Leben*, Verlag Chemie, Weinheim, 1949, p. 135.
(11) (a) Feyerabend, P. *Against Method*, New Left Books, London, 1975 (re-issued by Verso in 1993), p. 9–11. (b) For a breezy vignette of Feyerabend and his philosophy, see Horgan, J. *The End of Science*, Addison-Wesley, New York, 1996, p. 47ff.
(12) Mach, E. *Erkenntnis und Irrtum*, Neudruck, Wissenschaftliche Buchgesellschaft, Darmstadt, 1980, p. 200, as quoted in ref. 11.
(13) Yeats, W.B. *A Drinking Song*, from *Collected Poems of W.B. Yeats*, Finneran, R.J., ed. Macmillan, New York, 1983, p. 93.
(14) Holton, G. in *The Nature of Scientific Discovery*, A Symposium Commemorating the 500th Anniversary of the Birth of Nicolaus Copernicus, Gingerich, O., ed. Smithsonian Institution Press, Washington, 1975, pp. 216–217.

Chapter 2

Discoveries Missed, Discoveries Made: Two Case Studies of Creativity in Chemistry

2.1 Science and the Individual

Discovery in science evolves from the actions of people as conditioned by history and happenstance. By clarifying the influence of the personal, we can begin to see the origins of creativity. Perhaps, by thus sharpening our sensitivities, we may improve our own chances of becoming discoverers. Even if not, by scrutiny of the ideas, true and false, of our illustrious predecessors, we may at least enhance our own strengths and compensate for our weaknesses.

With these objectives in mind, I describe here two case histories of major discoveries in organic chemistry, the Diels-Alder "diene synthesis" and the Woodward-Hoffmann orbital symmetry rules. In these accounts, the missed opportunities have as much to teach us as the final successes.

2.2 Diels, Alder, Their Competitors, and the Discovery of the "Diene Synthesis" (Diels-Alder Reaction)

The formation of a cyclohexene by the conjugate addition of a 1,3-diene and an alkene (**1** + **2** → **3**, Scheme 1),[1] which has come to be called the diene synthesis or Diels-Alder reaction, has few rivals among the transformations of organic molecules for general applicability in synthesis and deep mechanistic implications.

Scheme 1

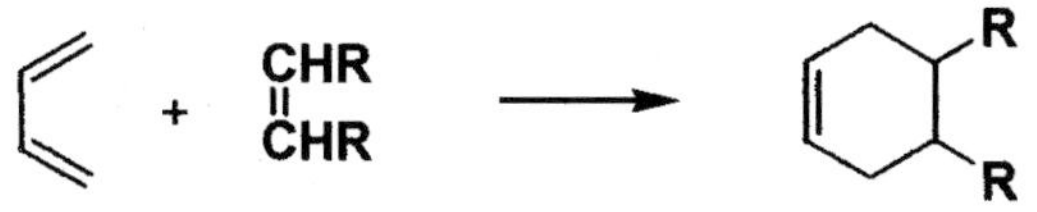

CHR
||
CHR

R
R

1 **2** **3**

At the outset, Otto Diels had no inkling of the existence of this reaction. In this sense, the eventual discovery was based upon a serendipitous observation, but to apply that word to the discovery as a whole conveys a faint odor of condescension and can be taken to imply that the discoverers stumbled all unaware upon it. This would be an unjust characterization of Diels's early work, which was guided at every step by clear analysis and careful experimentation.

The story begins in 1921, when Diels and Back,[2a] pursuing a line of investigation that had been initiated earlier by Curtius, published a paper on the reactions of diethyl azodicarboxylate **4**. Curtius had found that one could convert the ester functions to amide functions by reaction with primary amines (**4** → **5**, Scheme 2).

Scheme 2

$$EtO_2C{-}N{=}N{-}CO_2Et + 2\,RNH_2 \longrightarrow RHNOC{-}N{=}N{-}CONHR + 2\,EtOH$$

4 → **5**

Diels and Back found that although some aromatic amines conformed to this pattern of substitution, an exception emerged in the case of β-naphthylamine **6**, which gave an entirely different type of reaction, namely an addition leading to product **8** (Scheme 3).

Scheme 3

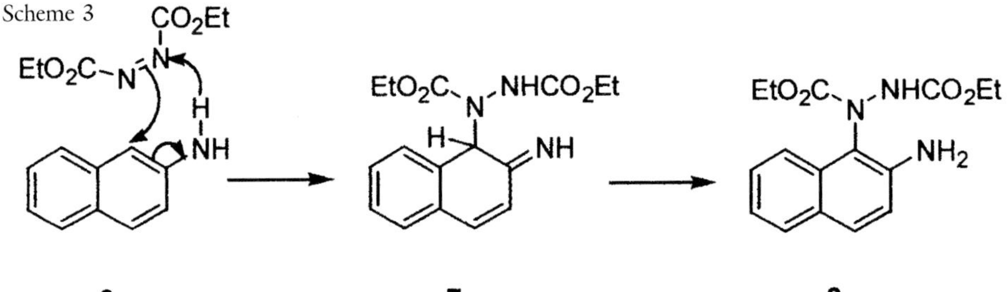

The nature of the reaction as an addition rather than a substitution was apparent just from the elemental analysis of the product, but Diels and Back expended a considerable effort to prove the structure by independent synthesis. The transformation involves overall the addition of a C–H bond of β-naphthylamine to the N=N bond of the azoester. Today we would think it likely that the mechanism passes over the enamine **7** via a process that is the formal equivalent of an ene reaction (although it may involve more than one step).

Now, I believe, came the crucial insight. Stimulated by these unexpected observations, Diels remembered that a 1906 paper by Albrecht[3] (see Section 2.3.3) had reported a formally similar type of reaction (Scheme 4) between quinone **9** and cyclopentadiene **10**, in which the overall change was thought to be addition of a

C–H bond of the diene to a C=C bond of the quinone to give the 1:1 adduct **11** and the 2:1 adduct **12**. This transformation, involving the C–H bond of a *hydrocarbon*, was most unusual, and Diels's ability to make the connection between two such apparently disparate pieces of information showed notable creativity.

Scheme 4

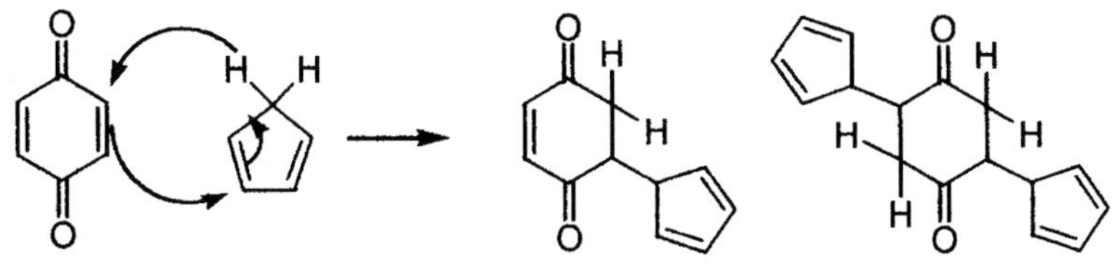

9 10 11 12

The obvious next step was to try the reaction of cyclopentadiene **10** with azodicarboxylic ester **4**. By analogy to the report of Albrecht, the now expected product should have the structure **13** (Scheme 5). However, as reported by Diels, Blom, and Koll,[2b] the actual product **14** was again of a new type, which resulted from a *cycloaddition* of C_1 and C_4 of the diene to the N=N bond.

Scheme 5

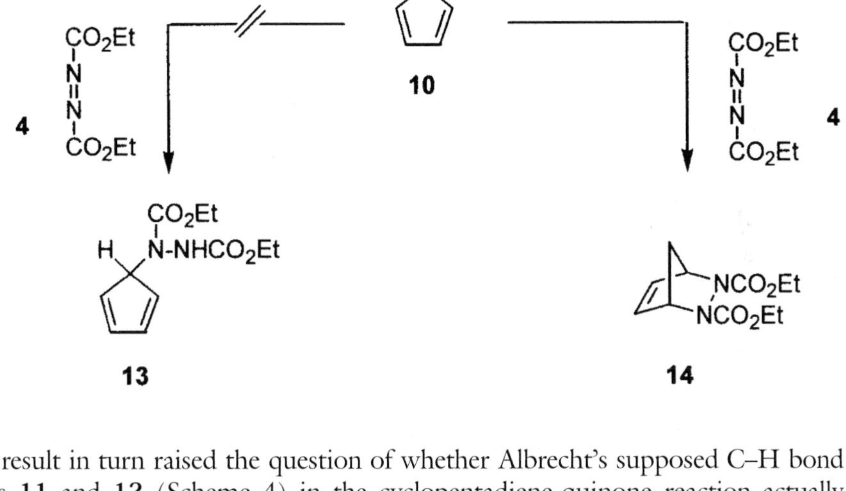

4 10 4 13 14

This result in turn raised the question of whether Albrecht's supposed C–H bond adducts **11** and **12** (Scheme 4) in the cyclopentadiene-quinone reaction actually might be the diene adducts **15** and **16** (Scheme 6).

Diels took up this problem with his student Kurt Alder.[2c] Their names were thereafter forever linked to this prototypical addition reaction when they showed beyond a shadow of doubt that Albrecht's 1:1 adduct and the corresponding 2:1 adduct were in fact diene addition products **15** and **16** (Scheme 6), respectively. The most telling observations were that the adducts **15** and **16** had only two C–C double bonds each, rather than the three and four, respectively, required by Albrecht's formulas **11** and **12** (Scheme 4).

Scheme 6

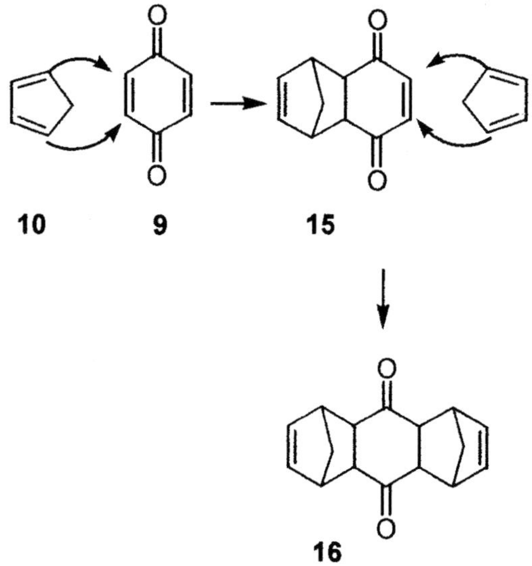

Diels and Alder themselves[2c] immediately grasped the significance of their work, both theoretical and synthetic:

» *Our results will play a role not only in the discussion of theoretically interesting questions, for example, the relationships of strain in polycyclic systems, but probably also will yield greater significance in a practical sense. Thus it appears to us that the possibility of synthesis of complex compounds related to or identical with natural products such as terpenes, sesquiterpenes, perhaps also alkaloids, has been moved to the near prospect.* «

With unmistakable severity, they issued a warning to would-be interlopers to stay out of the field they obviously now considered their own:

» *We explicitly reserve for ourselves the application of the reaction discovered by us to the solution of such problems.* «

Some inkling of the fierce territoriality they felt is evident in their chastisement[4] of Farmer and Warren,[5] who had had the temerity to question the [4 + 2] nature of the addition:

» *We take the view that the authors would have done better to have written one of us a private letter expressing their request and to have withheld the publication of their paper until the appearance of our thoroughgoing third contribution, which was then in press and has since appeared. Since we are engaged in filling the last holes in the understanding of the course of the diene synthesis, we direct to Messrs. Farmer and Warren the urgent request to disregard those events which obviously fall in the domain of the additon reactions we are studying.* «

Diels and Alder were right, of course, in the sense that their reaction did prove to be profoundly significant in the synthesis of natural products. It is just as well, however, that their claim of exclusive ownership came to be ignored by the rest of the chemical community, since they themselves made only minor contributions to natural product synthesis using their discovery. The reasons for this would take another paper to explore but probably include the distractions provided by the inexhaustible richness of the mechanistic and stereochemical questions associated with the diene addition, which occupied them in the 1930s and 1940s, Alder's discovery of the ene reaction, which diverted his attention even further, and the advent of World War II, which completely disrupted normal research in Germany. The first major moves toward the modern fulfillment of the Diels-Alder prediction came more than twenty years later, in in the early 1950s, with applications in the Woodward[6,7] and Sarett[8a] total syntheses of cortisone and in the Stork[8b] synthesis of cantharidin. In the Woodward design for cortisone **19** (Scheme 7), for example, even the casual reader could sense the arrival of new thinking about synthesis and could not escape the attraction of the compelling story that began with the first step, a Diels-Alder reaction between methoxytoluquinone **17** and butadiene **1** to give the adduct **18** (Scheme 7). The product incorporated functional groups for eventual elaboration to rings C and D of the cortisone structure.

Scheme 7

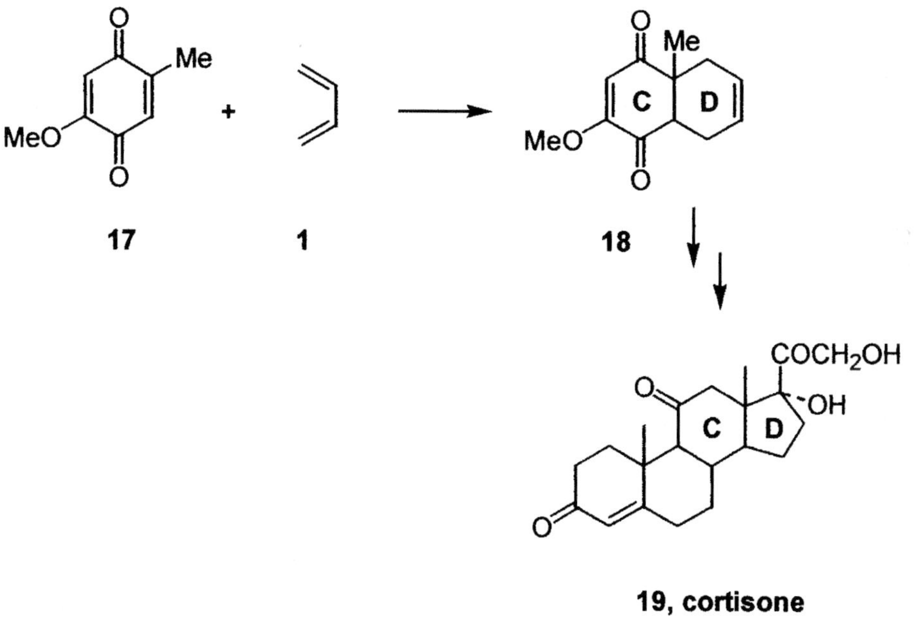

17 **1** **18**

19, cortisone

The profound significance of the diene synthesis was recognized by the award of the 1950 Nobel Prize in Chemistry jointly to Diels and Alder "for their discovery and development of the diene synthesis."

2.3 Predecessors and Near-Discoverers

Diels and Alder were not the first to observe a cyclohexene-forming diene-alkene cycloaddition. How did it happen that they emerged with the credit for this brilliant discovery? What I have to offer on this question depends heavily on conjecture, but there are a few clues. If we follow them, we may glimpse the interaction of forces and events that led to that outcome.

2.3.1 Early Workers

As early as 1892, Zincke had found[9a,b] and later correctly formulated[9c,d] a dimer of tetrachlorocyclopentadienone, and Lebedev[10] recognized vinylcyclohexene as the dimer of butadiene. It is perhaps not surprising that neither of these workers realized the generality of the process. Zincke's observation had been a side issue in his exploration of other unrelated phenomena. The reaction itself occurred under conditions in which the product was not stable, and its presence could be inferred only indirectly. Zincke did well to unravel a complicated story and had no obvious basis for discerning this particular golden nugget embedded in the hard rock of difficult but routine investigation.

Lebedev[10] was a participant in a controversy over the mechanism of formation of rubber (formally a polymer of isoprene) and other natural polymers. His attention and that of contemporaries such as Staudinger[11] and Harries[12] was directed elsewhere. (We should say "misdirected", since of course we now know, with the arrogance of our posteriority, that the ordinary thermal dimerization of aliphatic 1,3-dienes has nothing to do with the formation of natural rubber).

2.3.2 von Euler and a Near Miss

Well then, were there others in a better position to have discovered the [4 + 2] cycloadditon before Diels and Alder? The answer is yes. And here the determinative influence of personal circumstance begins to manifest itself. In Diels and Alder's first paper,[2c] they refer to the work of von Euler and Josephson,[13] who stated

»*In connection with not yet published studies on the course of the condensation of isoprene to rubber, one of us has approached more closely the question of how the group of atoms -CH=CH-CH=CH- reacts with other unsaturated groups, in which the goal of the study was, at least in favorable cases, a numerical value for the equilibrium [constant of the reaction 20 → 21, Scheme 8]. The necessary preliminary work has led to some new results of preparative significance.*«

Scheme 8

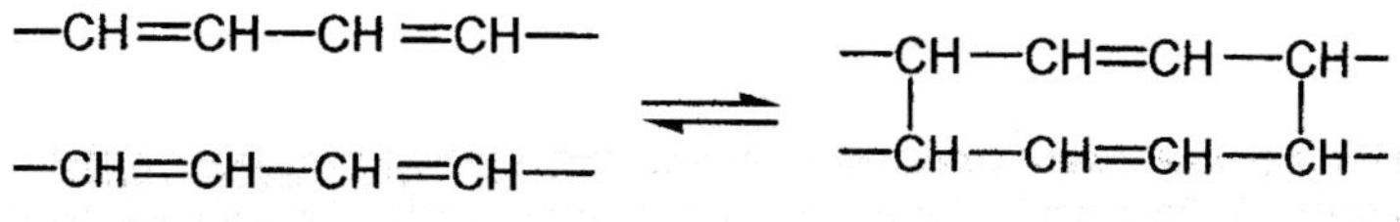

—CH=CH—CH=CH—
—CH=CH—CH=CH—

20

⇌

—CH—CH=CH—CH—
—CH—CH=CH—CH—

21

von Euler and Josephson proceeded to describe their observation of a 2:1 adduct of isoprene and p-quinone. Based essentially on two facts, namely the formation of a tetrabromide and a dioxime, they assigned the structure(s) **22** and **23** shown in Scheme 9. They realized that their job of structure proof was incomplete and talked their way out of it:

»*Because one of us had to interrupt the foregoing investigation, the reduction study has not yet been resumed; it will take place soon.*«

Scheme 9

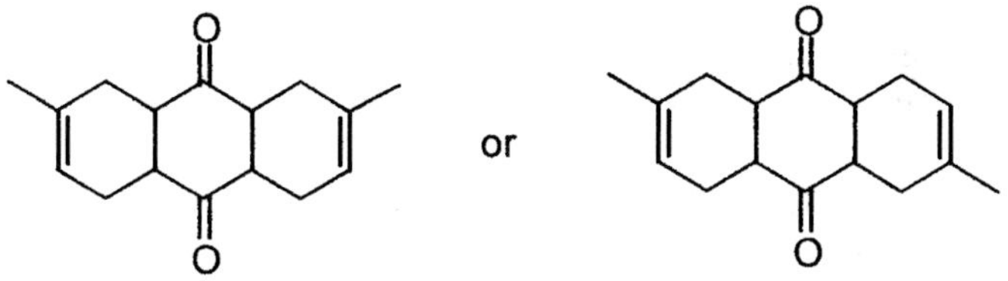

22 or **23**

This promise was never to be fulfilled. Clearly recognizing that they were onto something important, they pointed out that the structures they proposed

»*contain the assumption that each of the isoprene molecules uses one valence of the terminal carbon atoms, and both inside free valences saturate themselves by formation of a 2,3-double bond. This is completely in accord with the experience of conjugated double bonds.*«

At this point, it is difficult to suppress the feeling that von Euler, at the very threshhold of opportunity, turned and walked away. He and Josephson had the diene addition in their hands eight years before Diels and Alder! They had the correct structure of the adduct. They fully understood the mechanistic implication of their results as a manifestation of conjugate addition. And yet they never published another word on the subject. Why would a chemist give up a project of this significance apparently to pursue another task? Actually, who was Hans von Euler, and what was the other project that diverted his attention at this crucial moment?

From a biographical notice by Wilhelm Franke, *On Hans von Euler's 80th Birthday*,[14] we learn that he was (and had been since 1906, when he was 33 years old) Ordinarius Professor of General and Organic Chemistry at the University of Stockholm. Josephson was his student. Von Euler was to become director of the Biochemical Institute at that university in 1929, the same year that he shared the Nobel Prize in Chemistry (with A. Harden) for investigations on the fermentations of sugars and the fermentative enzymes.

Although he worked in Sweden for most of his life, von Euler was German – to the core. The son of a Bavarian military officer, he served Germany in both World Wars, in the first as a flyer and in the second with duties specified only as being "on the Turkish front." Franke[14] seems to suggest that von Euler's conduct in World War II was acceptable, since it was not politically motivated but rather based upon the moral stance of the soldier. Nevertheless, despite a number of expiatory actions by von Euler, many of his Swedish countrymen never forgave him his willing service to the German regime.

Be that as it may, there is no doubt that von Euler was a fabulous scientist, one of the most versatile and creative of the century. He was trained as a physical chemist, studying at one time or another with Landolt, Nernst, van't Hoff, and Arrhenius. He also became adept at organic chemistry as a result of brief but influential stays in the laboratories of Hantzsch and of Thiele (of whom we shall say much more in Sections 2.4 and 2.5). Probably under the influence of Arrhenius, von Euler became increasingly drawn toward biochemistry, and by 1910 was already started on the monumental series of studies of enzyme structure, kinetics, and mechanism that were to make him world-famous. What is astonishing is that while all this was going on, he was somehow able to find time for occasional forays into pure organic chemistry. This dazzling little thrust in 1920 with Josephson on the isoprene-quinone reaction was just one of many half-finished studies he seemed to toss off, as it were, almost for amusement.

We now can understand why von Euler could walk away from the discovery of the diene synthesis. He was deeply engaged in another massive project that had developed into his real life's work. It would have been irresponsible to permit a mere, although seductively promising, diversion to interfere.

Nevertheless, von Euler's ready use of the idea of conjugate addition stimulates another line of thought. Although the von Euler-Josephson paper does not refer to the source of information on what they called "the experience of conjugated double bonds", it seems certain that they had in mind the extensive studies by von Euler's former mentor Thiele in the 1890s, which had led to the idea of "partial valence". In fact, Thiele himself, the guru of conjugated systems, would have been a likely candidate to become the discoverer of the diene synthesis. How much of a leap of the imagination would it have required, after all, for the man who had established conjugate addition of dienes to several reagents (X–Y, Scheme 10) to realize that a similar addition to alkenes should occur? (see Scheme 11).

Scheme 10

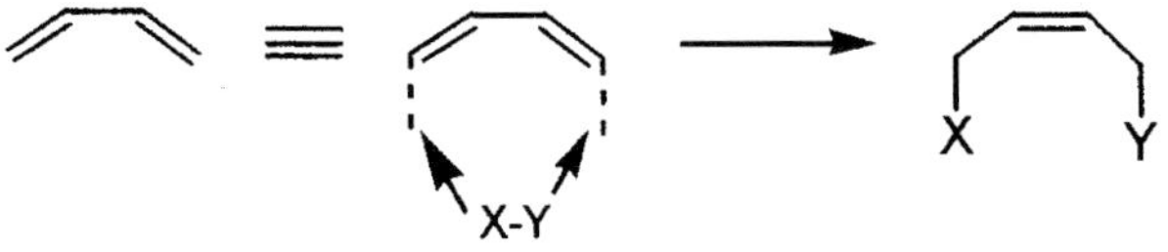

X-Y

X Y

1 24

Scheme 11

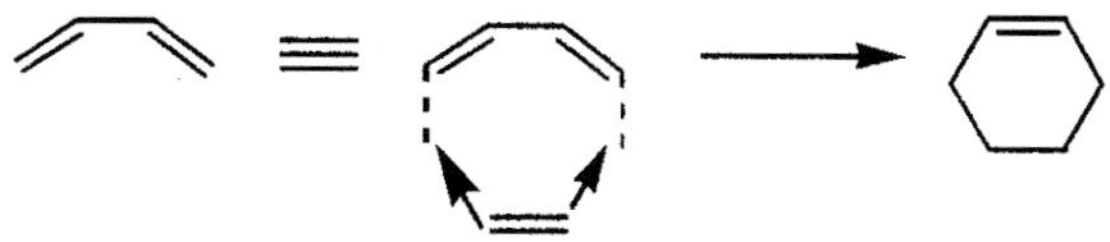

1 25

Although Diels and Alder's 1928 paper[2c] does not refer directly to Thiele's work as a forerunner of their own, they do emphasize that the diene addition amounts to reaction of the free valences at the 1,4-positions of the diene, an idea so widely associated with Thiele that they probably felt a reference unnecessary. So we have the inevitable question: Why did Thiele not discover the diene synthesis?

2.3.3 The Mystery of Albrecht

To approach the answer, we must start by examining more closely the paper of 1906 by Albrecht[3] which is reference 1 in the 1928 Diels-Alder paper. In fact, as we have seen, Albrecht's interpretation of his experimental results was a major target of the Diels-Alder paper. Albrecht had reported that the reaction of cyclopentadiene and quinone had given both a 1:1 and a 2:1 adduct. He was quite diffident about the structures of these and stated that the properties could not be reconciled with any probable structure, although he did write tentative formulas shown in Scheme 4. These represented the products to be expected if a C–H bond of cyclopentadiene added to a C=C double bond of quinone. As a result of the Diels-Alder work already described, Albrecht's structures ultimately had to be abandoned in favor of the bridged polycyclic structures **15** and **16** shown in Scheme 6.

In view of the moderate and almost apologetic claims Albrecht had made, the tone of the Diels-Alder criticism of his paper was harsher than one might have thought absolutely necessary. It may be that the territorial imperative was manifesting itself here also. In any case, a superficial reading of Albrecht's paper[3] leaves the impression that, with the handicap of the relatively undeveloped state of organic chemistry in 1906, he simply did not have the vision that Diels and Alder were able to bring to bear upon the problem 22 years later. But a second look reveals some puzzling questions.

Why had Albrecht been working on the reaction of cyclopentadiene and quinone in the first place? The introduction to the paper,[3] where one might have expected to find a stated reason (or at least a rationalization), is curiously blunt and uninformative. The first sentence reads merely: "Cyclopentadiene adds to one or two moles of quinone with extreme ease merely upon mixing, to give beautifully crystalline, stable compounds." Moreover, the heading of the paper gives no date of receipt of the manuscript and no indication of the laboratory where the work had been done. One might suppose that, given the rather relaxed editorial constraints of 1906, this omission was not totally startling, and that in the small chemical community of the time, the names of the major researchers alone were sufficient identification and carried with them the well known facts of their location. But Albrecht was not a major researcher. In the decade following 1906, Chemical Abstracts does not list a single publication under his name. So we have another question: Who was Walther Albrecht?

A strong hint of an answer is given by a footnoted reference in the Diels-Alder paper to Albrecht's "Inaugural Dissertation, München, 1902." Apparently, Albrecht had been a graduate student at Munich in the years just before 1902, and the work reported in his 1906 paper[3] must have been taken from his dissertation. [15,16] Oddly, Albrecht's paper gives no acknowledgment to his research advisor, but a moment's reflection leaves little doubt that it must have been ... Thiele!

2.4 Thiele and Albrecht

Johannes Thiele had come to Munich in 1893, and by 1900, the major part of his work on conjugate additions had been completed. Now he was heavily engaged in the synthesis and study of the fulvenes, which were highly reactive, colored compounds of great interest as potential synthetic building blocks and also as vehicles for the study of theories of color and constitution. Fulvenes **27** were made by the condensation of cyclopentadiene **10** with aldehydes or ketones **26**. In Scheme 12, I have formulated the condensation to emphasize that it is a dehydration, not an addition.

Scheme 12

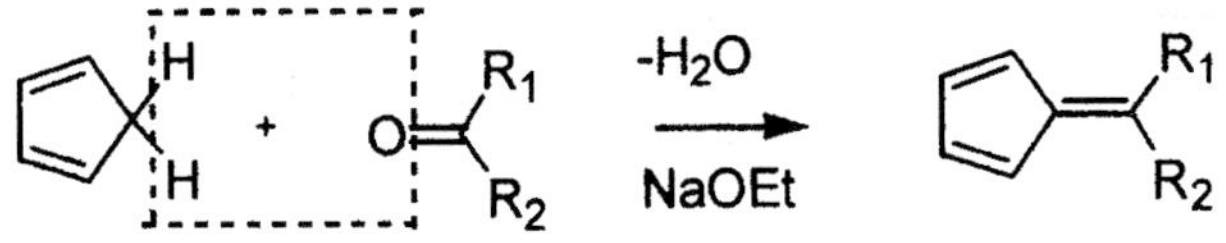

H + O R$_1$ R$_2$ H; $-H_2O$, NaOEt; R$_1$ R$_2$

10 **26** **27**

It is a reasonable conjecture that Thiele was fascinated with the prospect of synthesizing a double fulvene **28** by condensation of two moles of cyclopentadiene **10** with the conjugated diketone quinone **9** (Scheme 13):

Scheme 13

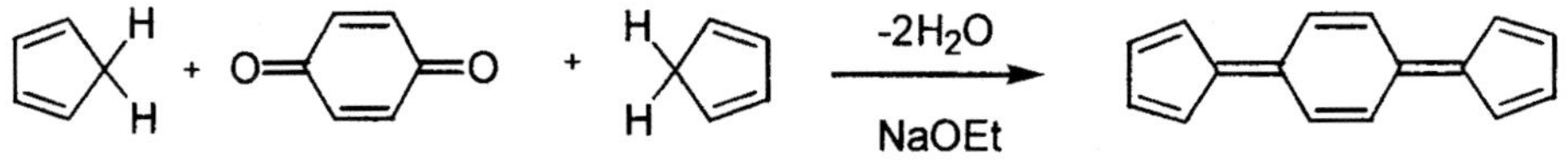

H H + O O + H H; $-2H_2O$, NaOEt

10 **9** **28**

None of this is forthcoming from Albrecht's laconic paper, but a hint that it is probably true comes from a paper by Thiele and Balhorn[17] in the very same issue of the Annalen. This article deals with the synthesis of several fulvenes. It does not refer to Albrecht's paper, but footnote 1 is an acknowledgment: "Several of the substances described in the following were prepared by Mr. W. Albrecht." The clinching evidence comes from Albrecht's dissertation itself,[15] the acknowledgment of which is to Thiele.

The first paragraph of the dissertation tells the whole story:

» *Thiele has condensed cyclopentadiene with aldehydes, ketones, and carboxylic esters to fulvene derivatives [Scheme 12] intensely colored compounds, which arise by elimination of water. Occupied with the elaboration of these compounds, I tested the behavior of cyclopentadiene toward quinones and found that benzoquinone reacts vigorously with the hydrocarbon. The two compounds combine without elimination of water.* «

The experimental section describes the conditions for carrying out the addition, which consist merely of mixing the components together neat, or more controllably, in an inert solvent. These conditions are quite different from the ones established by Thiele for making fulvenes, which always used sodium ethoxide as a "condensing agent." Today, we recognize that the function of the base was to convert cyclopentadiene to its anion. Although Albrecht's thesis contains no mention of it, I think it is highly likely that the base-mediated conditions must have been at least discussed by the professor and his student and probably tried in the cyclopentadiene-quinone reaction. It is my conjecture that this approach was abandoned because of the competing decomposition of quinone itself in basic medium. Given the primitive state of mechanistic understanding at the time, it would not be surprising if Albrecht had simply tried to carry out the reaction without the condensing agent. Once products had been obtained, the reigning theory of the laboratory, namely Thiele's ideas of partial valence, dominated the interpretation. Thus, quoting Thiele,[18] Albrecht argues[15]

» *In accord with its unsaturated nature, quinone is an extraordinarily reactive substance and is especially distinguished by its lively tendency to add hydrogen and substances with labile hydrogen. It contains a system of cross-conjugated double bonds. This leads to free partial valences not only at the terminal O-atoms, but also at the four inner C-atoms, so that it invites addition at not less than six points of attack, as is shown in the structural formula 29.*

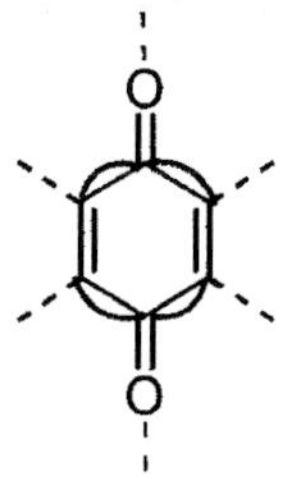

29

Cyclopentadiene condenses with aldehydes, ketones, and oxalic ester. The C=C double bonds evidently cause a loosening effect on the neighboring methylene group similar to that of C=O double bonds in ketones.

*The reactivity in condensation of indene **30** is somewhat diminished and that of fluorene **31** is significantly diminished relative to that of cyclopentadiene **10** (Scheme 14). Evidently the partial valences that influence the CH_2 group in cyclopentadiene are saturated within the attached benzene rings, thereby reducing the loosening effect on the methylene groups and the condensation reactivity.* «

Scheme 14

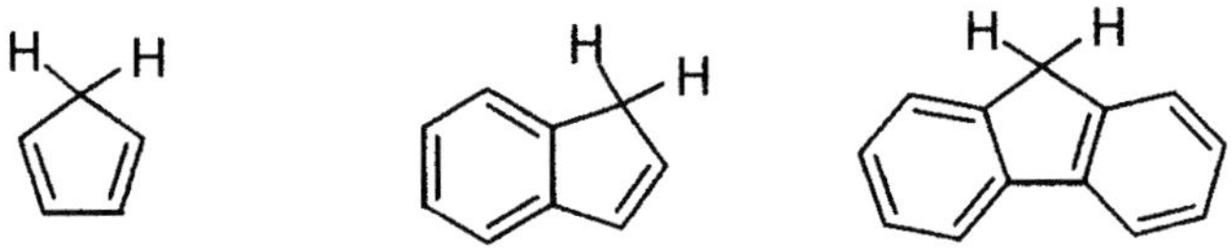

H H H H H H

10 30 31

cyclopentadiene **indene** **fluorene**

Of course, these ideas do not explain why cyclopentadiene reacts by *condensation* (with elimination of water) with other ketones but reacts by *addition* with quinone. It seems reasonable to speculate that Thiele was quite disappointed with the outcome of Albrecht's experiments. Thiele must have hoped for the "usual" cyclopentadiene condensation reaction, which would have given the spectacular double fulvene. I assume that when the results did not conform, Thiele wanted nothing more to do with the project and gave permission to Albrecht to publish the facts, but without Thiele's name on the paper. Apparently, Thiele in his own mind had gone beyond his conjugate addition phase; now he was thinking condensation. His failure to make this little mental jump back to addition reminds us of an iron law of science: a discovery is both a finding and a recognition of the finding. Thiele and Albrecht had fulfilled only half of the requirement. Thiele died in 1918 and so never had to endure the knowledge that the credit for what would have been the crowning manifestation of his partial valence theory ultimately would fall to others.

2.5 Thiele's Nature

Was Thiele's inaction here due to a momentary slip of attention, uncharacteristic of the man, or was his nature such that, great chemist though he was, these disappointments were likely, predictable, even fateful? We cannot be sure, but what his contemporaries write about him is at least consistent with the latter hypothesis. I offer thoughts on Thiele taken from the autobiography of Richard Willstatter.[19] A number of the factual details and some of the most important of the personal evaluations are corroborated in an obituary notice by Fritz Straus.[20]

In 1893, Adolf von Baeyer, leader of the chemical institute at the University of Munich, called Johannes Thiele to become the successor to E. Bamberger in the new position of Extraordinary Professor (Associate Professor). Thiele was only 29 years old and had recently completed his habilitation with Vollhard at Halle.

Despite his youth, Thiele already was known for his researches on guanidine derivatives. In one particularly startling paper, he had reported that the action of nitrous acid on aminoguanidinium nitrate **32** gave diazoguanidinium nitrate **33** (Scheme 15). Since aliphatic diazonium compounds were known to be extremely unstable and had not been isolated before, the result was remarkable. Unfortunately, the interpretation was shown much later by Hantzsch to be incorrect. Actually, the supposed diazonium nitrate **33** turned out to be a guanidinium azide **34**, and the reaction by which it was formed was an analog of the well known Curtius reaction in which acyl hydrazides upon nitrosation give acyl azides.

Scheme 15

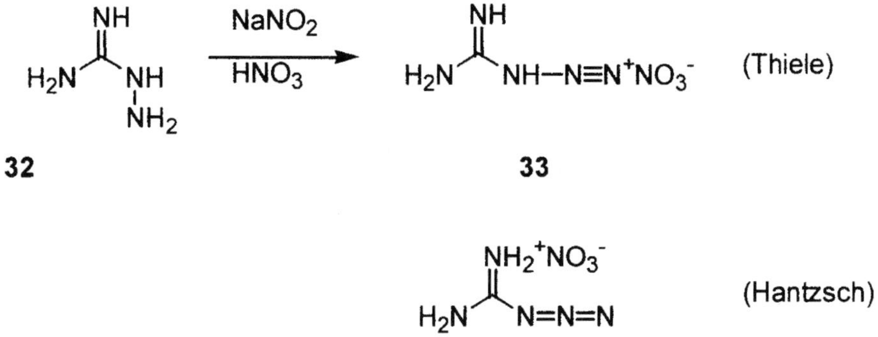

$$H_2N\text{–}C(=NH)\text{–}NH\text{–}NH_2 \ (\mathbf{32}) \xrightarrow[HNO_3]{NaNO_2} H_2N\text{–}C(=NH)\text{–}NH\text{–}N{\equiv}N^+NO_3^- \ (\mathbf{33}) \quad \text{(Thiele)}$$

$$H_2N\text{–}C(=NH_2^+NO_3^-)\text{–}N{=}N{=}N \quad \text{(Hantzsch)}$$

Willstätter expresses surprise that the "widely knowledgeable and keenly thoughtful" Thiele never did deduce the correct answer. It would be reasonable to attribute this mistake to Thiele's youth, and his failure to recognize it to his later preoccupation with other researches. However, it is also possible that we are seeing here part of a pattern of Thiele's inability to examine his own ideas critically and to realize that, if a mistake has been made, it must be corrected. As we shall see, Willstätter attributed to this intellectual flaw of Thiele's the blame for serious uncorrected errors in Thiele's later and far more significant work on conjugate addition. In Willstätter's words, "it seems to me that Thiele – in the style of the scientists after Baeyer – was more gifted to command than to listen."

In his early days at Munich, Thiele worked with vigor and determination. Willstätter says

»*The young Thiele was a powerful presence. He took over the position of head of the organic department, for which no tradition existed, and he made a model installation of it.*

Whereas the students had had no contact with his predecessor Bamberger, Thiele taught and ruled the laboratory and the student body with a new liveliness and sense of community. The combustion analyses were improved, the requirements in preparative training were raised, the students were stimulated and compelled to experiment with test tubes, dozens of which were kept clean in drawers and had to be filled in exactly prescribed ways. Doctoral candidates in greater numbers were graduated with well chosen and well guided dissertation projects. Military discipline ruled. And the students liked and respected this unfamiliar rule.

Thiele divided his time between the general laboratory and his small private laboratory. He visited not only his own co-workers but also, especially in the first years, every beginner. Also the lecture course, alternating great lectures on benzene derivatives and coal tar dyestuffs, adhered to a high level of instruction. «

Among Thiele's most important contributions were the series of researches on conjugate addition (see Scheme 10) which led to his theory of partial valence. The basic idea was that in a conjugated 1,3-diene, the valences at C-2 and C-3 would be only partially directed to their nominal partners at C-1 and C-4, respectively. The remaining partial valences at C-2 and C-3 would be directed toward each other, leaving partial valences unused at C-1 and C-4. Thus, addition of a reagent X–Y would occur preferentially at C-1 and C-4, and a double bond would appear between C-2 and C-3 (Scheme 16). As Willstätter points out, and as we discuss further below, Thiele pressed too hard on this point by insisting that the addition *must* occur 1,4.

Scheme 16

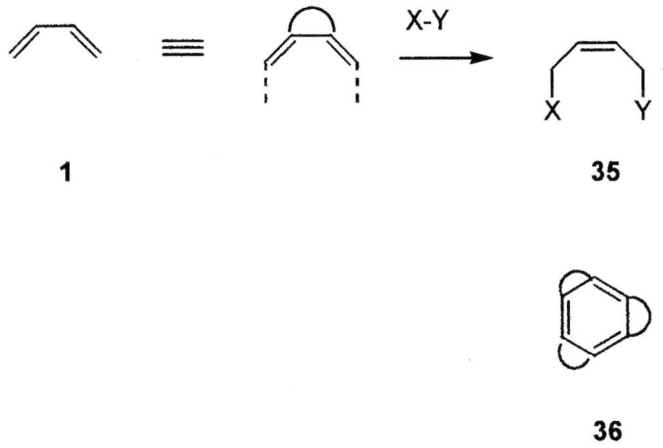

X-Y

X Y

1 **35**

36

An especially gratifying corollary of the theory was its explanation of aromatic character. As is symbolized in **36**, all of the the single bonds of the benzene nucleus are structurally analogous to the C-2-C-3 single bond of a 1,3-diene, and hence no partial valences are unused. Thus, benzene has "saturated" character despite the presence of its double bonds.

Thiele's work on conjugated systems made him famous in his time. His theory of partial valence was the most sophisticated of the pre-quantum mechanical attempts to understand the relationship between structure and reactivity. Eventually, of course, we must agree with the evaluation of Walter Hückel:[21]

»*While it is quite true that today this hypothesis must be regarded as obsolete and superseded ... [it] did nevertheless enable us to foresee somewhat, or hazard a guess, as to what has today, though in a somewhat changed form, assumed real shape.*«

In Willstätter's account,[19] Thiele changed rapidly, becoming obese and sluggish, and losing interest in his research and his students. Thiele was 37 years old when he received a call to Strasbourg, which he delightedly accepted, but by that time, his best work had been done.

Willstätter speculates on the possible reasons for Thiele's decline:

»*Was the cause of his alteration that Thiele lacked the strength to correct his mistakes, or did the weakness of this strong man, that he could not admit error, bring about the early conclusion of his scientific development? Toward the end of his time in Munich, Thiele encountered important examples of additions that did not follow his rule of 1,4-addition and in fact contradicted his published experimental statements. These results were difficult for Thiele to bear. They were never published.*«

Although his personality and background would not naturally have drawn him close to Thiele, Willstätter clearly felt tremendous respect and admiration for his colleague. Nevertheless, Willstätter cast a cold eye on those of Thiele's deficiencies that ultimately arrested his scientific growth. One wonders whether it would have been in vain to offer these criticisms to the suffering Thiele directly. Would they have been received with enough objectivity to have benefited him? Probably not. Few of us have the courage or self-confidence to give or to receive such advice. But the insights into Thiele's character we have just read support the idea that his formulaic, systematic way of thinking about science was not the right intellectual equipment to cope effectively with the unexpected. Doing science at the highest, most creative level is a bit like playing jazz. One has to be disciplined, yes, but also open, flexible, ready to abandon an idea and pick up a new one when Nature thrusts it upon one.

As luck would have it, Willstätter's book contains a remarkable photograph (Figure 1) of the Munich organic chemistry research group in 1901. In addition to the great Baeyer himself, surrounded by his associates, it shows several other members of the faculty, including Koenigs, Einhorn, Willstätter, and Thiele. There also is a picture of the young Jakob Meisenheimer, a student of Thiele, who was to become a renowned professor in his own right. Several names are listed without initials. No doubt these are graduate students, a conjecture that receives some support from their modest placement in the rear of this distinguished company. Among them we find the once-mysterious Albrecht. Given the timing of this event, it seems likely that when the photography session ended, he may well have returned directly to his lab bench to continue his work on the cyclopentadiene-quinone reaction.

Were this a world of pure logic, Thiele rightfully should have been the discoverer of the diene synthesis. There even seems to be irony in the near–homophonous

Figure 1. A photograph of research workers at the Chemical Institute of the University of Munich in 1901. Professor Baeyer sits in the front row holding his hat. On his right is Professor Koenigs, and seated on his immediate left is Professor Thiele. Professor Einhorn is on Thiele's left. Willstätter is the bearded man standing just beyond Baeyer's right shoulder, and to his immediate right, in a lab apron, is Meisenheimer. Albrecht is the clean-shaven young man with a bow tie, hair parted in the middle, two rows straight back from Willstätter. Taken from Table VII of Willstätter's book (reference 19) and reproduced with permission of Verlag Chemie.

names of the discoverers and the near-discoverers: Diels-Alder/Thiele-Albrecht. Probably we ask too much if we insist that when his window of opportunity opened in 1902, Thiele should have made the mental connection that would have been necessary. Unusual insight would have been required to do that so early. On the other hand, as we have seen, there was a stiffness in his thinking that might have prevented him from ever seeing the answer. Perhaps it is fair to say of this, with Cassius, "the fault, dear Brutus, is not in our stars, but in ourselves."

2.6 The Alternation Effect and the Discovery of Orbital Symmetry Conservation

In 1981, the Royal Swedish Academy of Sciences, recognizing work carried out in the 1950s and 1960s, awarded Nobel Prizes in Chemistry to Kenichi Fukui of Kyoto University and Roald Hoffmann of Cornell University "for their theories, developed independently, concerning the course of chemical reactions." Hoffmann had collaborated closely with Robert Burns Woodward of Harvard University. Although Woodward already was a Nobel Laureate for other work, few chemists doubt that had he lived, his contributions to this field would have placed him again in Stockholm, alongside Fukui and Hoffmann.

2.7 Experimental Stimulus for the Orbital Symmetry Rules

The Nobel citations mention "theories," but it is well to keep in mind that, at least the Woodward-Hoffmann work was driven in the first instance by experiment. No one tells this story as well as Woodward himself. His account,[22a] rich in characteristic darts, feints, digressions, paradoxes, puzzles, and jocularities, is too long for us to repeat here, but we can focus on the original motivation, which was the synthesis of a dauntingly complex molecule, Vitamin B_{12}. Woodward describes the observation of the peculiar pattern of behavior of the synthetic intermediates **37–40** shown here (in partial structures, Scheme 17):

Scheme 17

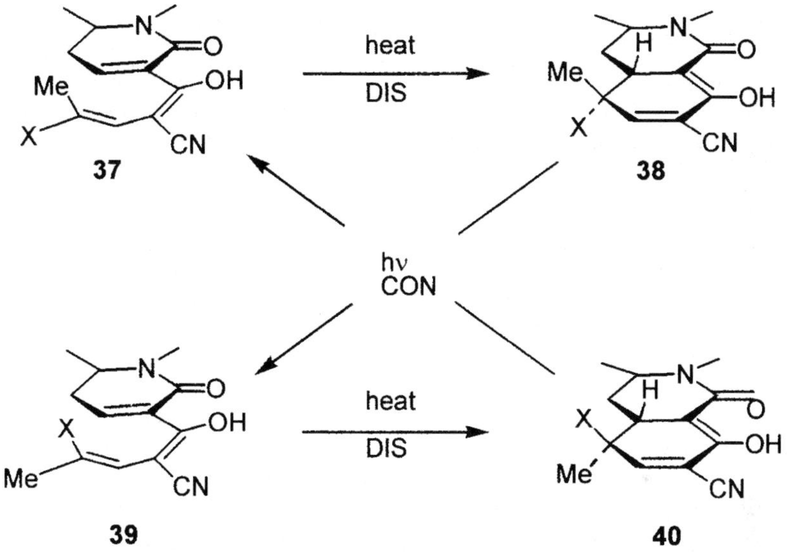

Note that the ring-closure or opening is disrotatory (rotational displacements at the breaking or forming bond in opposite directions) in the thermal reactions and conrotatory (rotational displacements at the breaking or forming bond in the same direction) in the photochemical reactions. Moreover, Woodward and Hoffmann recalled that some years before, Vogel[23] had reported the highly stereospecific thermal ring-opening of dimethyl cyclobutene-*cis*-3,4-dicarboxylate to dimethyl *cis*, *trans*-muconate (**41** → **42**, Scheme 18, X = CO_2CH_3).

Scheme 18

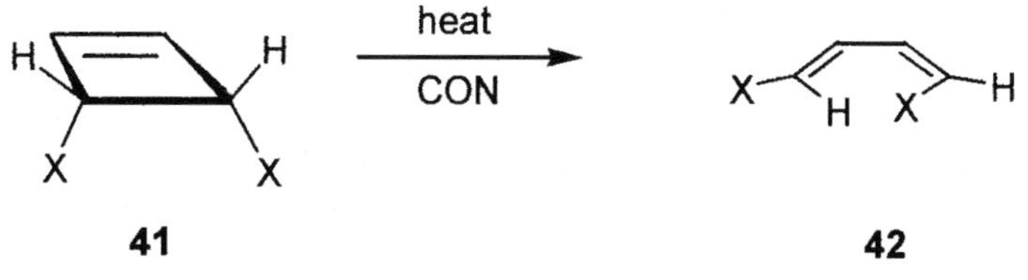

Now here comes a crucial insight. The thermal cyclization of the triene system in the Vitamin B12 synthesis is cleanly *disrotatory*, but the thermal (de)cyclization in Vogel's compound is cleanly *conrotatory*. Woodward says[22a]:

» *Here then we were faced with a most remarkable situation; two obviously closely related processes, each of them occurring with essentially complete stereospecificity, but in each case the stereospecificity was precisely the opposite, that is, in the cyclobutene opening the conrotatory mode was followed, while in the formation of cyclohexadiene from the open-chain triene a disrotatory process was cleanly followed. It now seemed certain to us that there must be a very fundamental factor at work in these reactions, and it was not long before we discerned that there might be a correlation between these facts and the symmetries of the molecular orbitals of the π-systems involved.* «

2.8 Theory

The experimental observation is that there is an alternation of the conrotatory and disrotatory modes, with the determining factor being the number of π-electron centers in the linear polyene chain. In thermal reactions, four such centers give conrotation and six give disrotation. The basic theoretical idea is that this alternation is caused by the alternation of the symmetry and antisymmetry of the molecular orbitals of linear polyenes. The stereochemical course of the reaction is held to be controlled by the symmetry or antisymmetry of the highest occupied molecular orbital.

This also offers an explanation for the switch from disrotatory thermal reaction to conrotatory photochemical reaction observed in the triene system of the Vitamin B_{12} intermediate. On the assumption that the first excited state of a photochemical reactant is the one actually undergoing reaction, the highest occupied orbital would be of opposite symmetry to that of the ground state molecule, and a simple extension of the rule would predict exactly the observed change.[24a]

These ideas were first put into print in a series of short communications in 1965.[22] In bold and sweeping extrapolations, Woodward and Hoffmann elaborated them to cover a wide range of other kinds of so-called concerted pericyclic reactions, including many that had not yet been observed. Experimental evidence supporting many of the predictions soon came, and this kind of theoretical analysis became commonplace in the literature.

2.9 Predecessors: Havinga, Schlatmann, Oosterhoff

Nevertheless, the alternation effect that produces opposite stereospecificities for thermal and photochemical cyclizations of linear polyenes had been observed prior to Woodward's Vitamin B_{12} work, as Woodward and Hoffmann pointed out.[22b] In carefully documented studies by Havinga and Schlatmann at Leiden, reported in

1961[25], the triene **47** exhibited exactly the same preference for photochemical conrotation and thermal disrotation (Scheme 19).

Scheme 19

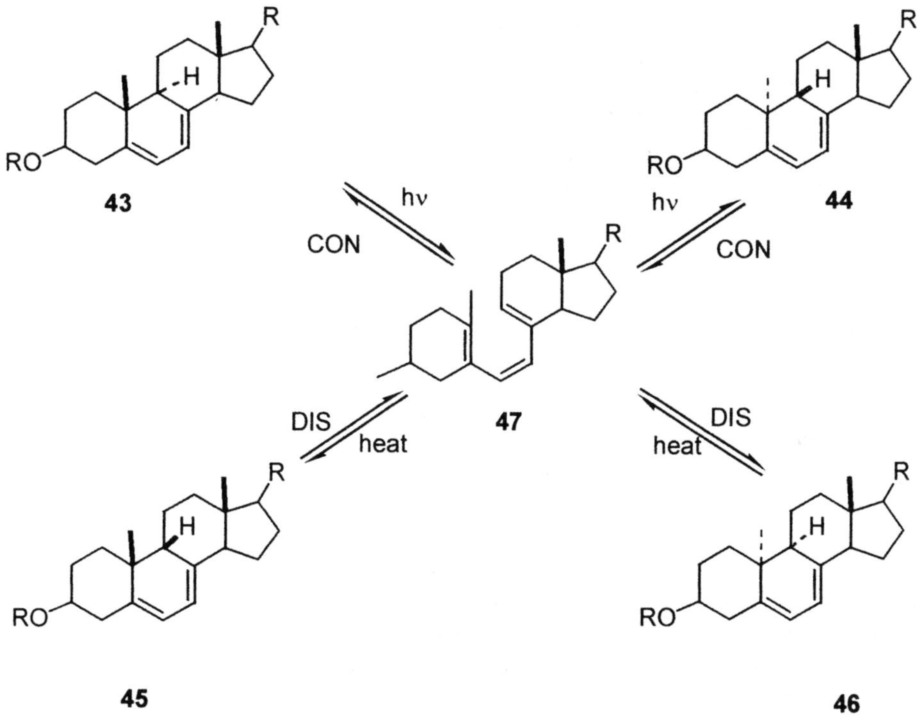

RO R H **43**
RO R H **44**
hν CON
hν CON
R **47**
DIS heat
heat DIS
RO R H **45**
RO R H **46**

Havinga and Schlatmann considered several explanations of this behavior but did not arrive at a definite hypothesis. Two sentences near the end of the paper, however, refer to personal discussions with Oosterhoff. These took on great significance in the aftermath of the Woodward-Hoffmann orbital symmetry excitement four years later:

» *As Prof. Oosterhoff pointed out, another factor that possibly contributes to the stereochemical difference between the thermal and photo-induced ring closure may be found in the symmetry characteristics of the highest occupied π-orbital of the conjugated hexatriene system. In the photo excited state this highest occupied orbital is antisymmetric with regard to the plane that is perpendicular to the bond 6,7 making* syn *approach less favourable.* «

Figure 2. Luitzen J. Oosterhoff (1907–1974), Professor of Theoretical Organic Chemistry, University of Leiden. Taken from *Enjoying Organic Chemistry*, Havinga, E. from *Profiles, Pathways, and Dreams*, Seeman, J.I., ed. American Chemical Society, Washington, D.C. 1991, p. 35. The image is reproduced with the kind permission of Mrs. Ulrika W. Brongersma-Oosterhoff.

2.10 Oosterhoff's Scientific Style

Luitzen J. Oosterhoff (1907–1974) (see Figure 2) was an outstanding theoretical organic chemist at Leiden, a logical person to have solved this problem. In fact, in a sense it might be said that he did solve it, for his explanation, although apparently offerred with considerable diffidence, was essentially the same as that of Woodward and Hoffmann's main idea. Why then did Oosterhoff not follow up on his insight? Why did he not generalize from it and cause a revolution in the way chemists think about reaction mechanisms?

I conjecture that the reason was that Oosterhoff thought that the explanation was oversimplified if not actually wrong.[26] He said as much in a conversation at the Bürgenstock Conference on Stereochemistry in 1966. With his sophistication about photochemistry, he recognized that it was by no means obvious that the photochemically active state in these reactions was the first excited state, as Woodward and Hoffmann had postulated. To him, it seemed likely that the mechanism was more complicated, especially since, as Dauben[27] had pointed out, the ground state of cyclobutene is about 20 kcal/mole less stable than that of butadiene, and therefore the spectroscopic singlet of cyclobutene is about 50–60 kcal/mole higher than that for butadiene. A direct transformation of the butadiene first excited state into the cyclobutene first excited state, as envisioned by Woodward and Hoffmann, therefore would be energetically improbable.

Oosterhoff worried and worried at this question, and in 1968, he and his student van der Lugt produced a paper[28] based upon valence bond theoretical calculations that gave a definitive answer. Their results are summarized in the state diagram (Figure 3). This demonstrates the energies of the ground state and the first two excited states of butadiene and cyclobutene as functions of an angle, whose value is a shorthand way of tracking the extent of progress along the reaction coordinate. Note that in accord with the Woodward-Hoffmann rules, the ground state strongly prefers the conrotatory pathway. The lowest energy photoexcited state at the planar butadiene geometry (Franck-Condon transition) is the antisymmetric one, but in the disrotatory pathway headed toward cyclobutene, this state makes an "intended" crossing with a higher-lying but rapidly descending symmetric excited state. In nuclear configurations lacking two-fold symmetry, these states mix, and the crossing is avoided, leading to emergence of the symmetric state at lower energy than the antisymmetric one. This rather deep well or funnel occurs (necessarily, according to Oosterhoff's theory) close in geometry to the ground state energy maximum characteristic of a Woodward-Hoffmann "forbidden" reaction. Passage then to the symmetric ground state surface occurs by a radiationless process. The overall result is that disrotation is preferred, in agreement with experiment,[29] but the excited state involved is not that assumed in the original orbital symmetry formulation.

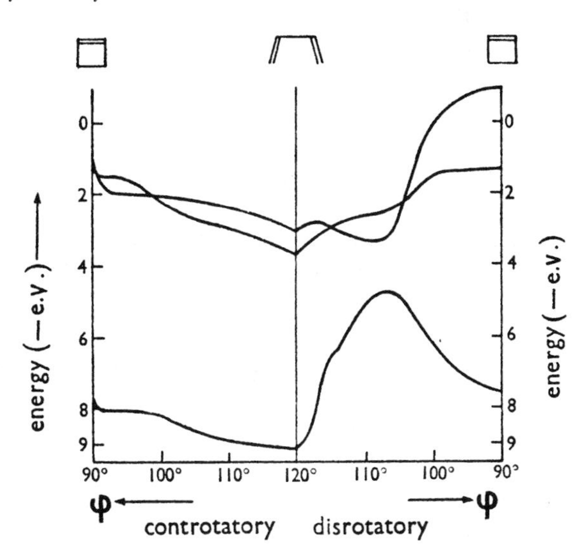

Figure 3. Energies of the ground state and two excited states during the cyclization of 1,3-butadiene to cyclobutene. Taken from reference 28 and reproduced with permission of the Royal Society of Chemistry.

It is entirely understandable that a mind as penetrating as Oosterhoff's could not let go of this challenging problem until a solution had been found. The beauty and depth of his analysis are undeniable. Moreover, the idea of a reaction funnel created by close energetic and geometric approach of an excited state minimum and a ground state maximum has great significance for the understanding of photochemical reactions. Yet it must be said that Oosterhoff, in pursuing this puzzle to its resolution, allowed himself to be diverted from the even more significant implications of the alternation effect.

2.11 Conclusions

On the way to the discovery of the diene-alkene cycloaddition, as we have seen, the two nearest misses before Diels and Alder were those of von Euler and of Thiele. In my view, von Euler's failure to exploit his semi-discovery resulted not from defects of mind, but rather from the circumstances of his personal commitment to his enzyme research at that crucial moment of his career. On the other hand, I think it was Thiele's intellectual rigidity and defensiveness that would not permit him to take the one little step he needed to add the diene reaction to the list of conjugate additions he already had discovered.

In the case of the electrocyclic reactions, Oosterhoff's keen insight and mental toughness drove him to the solution of the photochemical funnel problem but forbade him from making the broad but (in his view) unacceptably approximate generalizations that would have led to the orbital symmetry theory.

Putting aside the circumstantial differences of the two cases, and without intending disrespect to Thiele and Oosterhoff, who were both great chemists, one might say that the near-misses in the recognition of the diene synthesis and of the orbital symmetry theory had in common a certain failure of the imagination.

The lessons for the working scientist in these stories are ones we already know, whether or not we have articulated them to ourselves: The first, dramatically illustrated by the von Euler story, is that, in the progress of science, as in life itself, all things are subject, in Shelley's words,[30] to "fate, time, occasion, chance and change." The second, which emerges from the Thiele and Oosterhoff stories, is that the defining property, which is also the immutable precondition, of creativity is the proper mixture of rational and imaginative elements. In the ancient intellectual polarization that goes back to the apollonian and dionysian ideals, one cannot choose sides. One needs the right blend of order, discipline, and clarity, with ardor and unfettered fantasy. Perhaps even the vast majority of us whose creative powers are modest can benefit if we recognize the need for this balance.

2.12 Acknowledgments

This work originally was prepared for presentation as one of the Ralph Hirschmann Lectures at Oberlin College and as one of the Reynold C. Fuson Lectures at the University of Nevada, Reno, in April, 1991. A slightly different version has appeared in print: Berson, J.A. *Tetrahedron*, **1992**, *48*, 3. I thank Professors J. Cornelisse, M. Gates, R. Hoffmann, H.C. Longuet-Higgins, K. Mislow, and H. H. Wasserman for helpful comments.

2.13 References

(1) For a review that covers the early history of the reaction, see Kloetzel, M.C. *Org. Reactions*, **1950,** *4*, 1.

(2) (a) Diels, O.; Back, J. *Chem. Ber.* **1921,** *54*, 213. (b) Diels,O.; Blom, J.H.; Koll, W. *Ann.* **1925,** *443*, 242. (c) Diels, O.; Alder, K. *Ann.* **1928,** *460*, 98.

(3) Albrecht, W. *Ann.* **1906,** *348*, 31.

(4) Diels, O.; Alder, K. *Ber.* **1929,** *62*, 2087.

(5) Farmer, E.H.; Warren, F.L. *J. Chem. Soc.*, **1929,** 897.

(6) Woodward, R.B.; Sondheimer, F.; Taub, D.; Heusler, K.; McLamore, W.M. *J. Am. Chem. Soc*, **1951,** *73*, 2403.

(7) Actually, the first attempt to effect the synthesis of steroids using Diels-Alder chemistry seems to have occurred just before World War II, in the work of Dane and Schmitt, which is further described in Chapter 4. See Dane, E.; Schmitt, J. *Ann.* **1939,** *537*, 246.

(8) (a) Sarett, L.H.; Arth, G.E.; Lukes,R.M.; Beyler, R.E.; Poos, G.I.; Johns, W.F.; Constantin, J.M. *J. Am. Chem. Soc.*, **1952,** *74*, 4974. (b) Stork, G.; van Tamelen, E.E.; Friedman, L.J.; Burgstahler, A.W. *J. Am.Chem. Soc.* **1951,** *73*, 4501.

(9) (a) Zincke, T.; Günther, H. *Ann.* **1892,** *272*, 243. (b) Zincke, T; Bergmann, F.; Francke, B.; Prenntzell, W. *Ann.* **297,** *296*, 135 (1897). (c) Zincke, T.; Meyer, K.H. *Ann.* **1909,** *367*, 1. (d) Zincke, T.; Pfaffendorf, W. *Ann.*, **1912,** *394*, 3.

(10) (a) Lebedev, S.V. *J. Russ. Phys. Chem. Soc.*, **1910,** *42*, 949. (b) *Chem. Abstr.* **1912,** *6*, 2009.

(11) Staudinger, H. *Die Ketene*, F. Enke, Stuttgart, 1912, p. 59, fn. 2.

(12) Harries, C. *Ber.* **1905,** *38*, 1195.

(13) von Euler, H.; Josephson, K.O. *Ber* **1920,** *53*, 822.

(14) Franke, W. *Naturwissenschaften*, **1953,** *40*, 177 (1953).

(15) Albrecht, W. *Über Cyclopentadienchinone*. Inaugural Dissertation, Munich (1902).

(16) I am grateful for a photocopy which was made available through the cooperation of the Kline Science Library at Yale and the New York Public Library.

(17) Thiele, J.; Balhorn, H. *Ann.* **1906,** *348*.

(18) Thiele, J. *Ann.* **1898,** *306*, 132.

(19) Willstätter, R. *Aus Meinem Leben*, Verlag Chemie, Weinheim, 1949, p. 59.(freely translated).

(20) Straus, F. *Z. Angew. Chem.*, **1918,** *31*, 117.

(21) Hückel, W. *Theoretical Principles of Organic Chemistry*, Elsevier, 1955, Vol. I, p. 693.

(22) (a) Woodward, R.B. in *Aromaticity*, Chemical Society Special Publication No. 21, London, 1967, p. 217. (b) Woodward, R.B.; Hoffmann, R. *J. Am. Chem. Soc.*, **1965,** *87*, 395. (c) *ibid.* **1965,** *87*, 2511. (d) Hoffmann, R.; Woodward, R.B. *ibid.* **1965,** *87*, 2046. (e) Subsequently, as a result of work by Fukui, Longuet-Higgins, Zimmerman, and Dewar, among others, a number of alternative ways to derive the same results came to light. For references, see ref. 24b.
(23) Vogel, E. *Ann.* **1958,** *615*, 14.
(24) Woodward, R.B.; Hoffmann, R. *The Conservation of Orbital Symmetry*, Academic Press, New York, 1970. (a) p. 45. (b) p. 176.
(25) E. Havinga, E.; Schlatmann, J.L.M.A. *Tetrahedron*, **1961,** *16*, 146.
(26) An interesting light on Oosterhoff's values in science is given in an article by Mulder, J.C.C. *Chemisch Weekblad*, 1990, 34, August 23, 1990. I am indebted to R. Hoffmann for this reference.
(27) Dauben, W.G. 13th Chemistry Conference of the Solvay Institute, *Reactivity of the Photoexcited Organic Molecule*, Interscience, New York, 1967, p. 171.
(28) van der Lugt, W. Th. A. M.; Oosterhoff, L.J. *Chem. Communs.* **1968,** 1235.
(29) A simple example is the photocyclization of *trans, trans*-2,4-hexadiene to *cis*-3,4-dimethylcyclobutene: Srinivasan, R. *J. Am. Chem. Soc.*, **1968,** *90*, 4498.
(30) Shelley, P.B., *Prometheus Unbound*, Act II, Scene IV, line 115 ff. Modern Library, New York, NY, undated, p. 261. The full text is " ... Fate, Time, Occasion, Chance and Change ... to these all things are subject but eternal love." I am indebted to Ms. M. Powell, librarian, Yale University Library, for directing me to the origin of this passage.

Chapter 3

Erich Hückel and the Theory of Aromaticity: Reflections on Theory and Experiment

3.1 Hückel's Contributions

In the period between 1930 and 1937, Erich Hückel (Figure 1), a theoretical physicist, made profound contributions to organic chemistry in his quantum mechanical descriptions of unsaturated and conjugated compounds. Although his work eventually proved influential, his career path was far from smooth, and for much of his life, he remained estranged from later developments that were based firmly upon his discoveries. The story of how this came to be the case, and how he won final vindication, vividly confirms the admonition that it is usually unwise to suppose that you can let your work speak for itself.

Figure 1. Erich Hückel (1896–1980), Professor of Theoretical Physics, University of Marburg. The image is reproduced from ref. 1 with permission of VCH Publishers. Photo Tita Binz, Mannheim.

This chapter gives a brief account of his academic career and simplified expositions, from the point of view of an organic chemist, of the highly approximate quantum mechanical methods he used to develop fruitful descriptions of olefinic and aromatic molecules. Of special significance in the case of cyclic molecules of the class C_nH_n is the concept of filled shells when the number of π-electrons is $(4N + 2)$ ($N = 0, 1, 2, \ldots$). Examinations of key explicative and predictive applications of these ideas reveal how they eventually motivated the exploration of new fields in organic synthesis and mechanism. Another significant (but until recently, virtually ignored) contribution by Hückel was the recognition that atomic connectivity is a strong determinant of electronic multiplicity in non-Kekulé molecules. This idea provides in principle a basis for predicting violations of Hund's rule, as recent computational and experimental developments confirm.

Also, we consider the interaction of experiment with different styles of quantum theory, and the impact of this relationship on the development of chemistry. As we shall see, Hückel's work exerted little influence on organic chemistry for decades before its importance finally began to be recognized. Moreover, although Hückel published his innovative ideas early in his career, after the age of 40 he made little further contribution to the field he had created. By attempting to determine the reasons for this truncation of his development and the for the reluctance of the chemical community to adopt his ideas, we may learn something of the complex but usually hidden forces that influence the growth of a scientific discipline.

3.2 Biographical Introduction[1–3]

Erich Hückel was born on August 9, 1896, the second of three sons of Marie and Armand Hückel. The intellectual development of the three boys was strongly influenced by their father, a physician with an interest in pure science. In the family line were several scientific notables, including the distinguished botanist Josef Gärtner (1732–1791). Erich Hückel enriched this heritage in 1921, when he married Anne Zsigmondy, the daughter of Professor Richard Zsigmondy of Göttingen, a renowned colloid chemist (Nobel Prize in Chemistry, 1925). The marriage ultimately was blessed with four children, three sons and a daughter, and endured until Hückel's death in 1980. Erich Hückel's brothers, Rudi (1899–1949) and Walter (1895–1973), followed in the family tradition. Rudi became a physician, but died prematurely. Walter achieved prominence as a professor of organic chemistry and a prolific author of significant research contributions and influential textbooks.[4]

Erich Hückel was educated as a physicist. His thesis for the D. Phil. at Göttingen (1921) under the supervision of Peter Debye was an experimental study of the scattering of X-rays. Afterward, he served briefly as an assistant to the mathematician David Hilbert and the physicist Max Born at Göttingen but then rejoined Debye, who had moved to the Eidgenössische Technische Hochschule in Zürich. Hückel stayed in Zürich until 1927. Two years on a Rockefeller Foundation scholarship, which he spent with Donnan in London and with Bohr in Copenhagen, were followed by two more years on a Deutsche Notgemeinschaft scholarship in Leipzig with Heisenberg and Hund. It was in Leipzig in 1930 that Hückel finished the first of his landmark papers on organic quantum chemistry.

Note that nine years after the D. Phil, Hückel still had no permanent job. In effect, his situation was like that of all too many of today's postdoctorals, who float in semi-permanent, semi-employed limbo. Through Debye's intervention, a *dozentur* of sorts was arranged for him at the Technische Hochschule in Stuttgart during the period 1930–1937. However, this was not a regularly budgeted position, and even Hückel's salary was insecure. Writing in 1979,[1] Hückel tells of how his wife often referred to this time as "seven years of disgrace." In 1966, nearly 30 years after his departure, the Technische Hochschule in Stuttgart was to recognize Hückel's qualities with an honorary degree.

Finally, in 1937 came a call to Marburg as Extraordinary Professor (Associate Professor) of Theoretical Physics. For a few months spanning 1945–1946, immediately following the end of the European phase of World War II, Hückel was separated from his job by the Allied authorities because of his pre-war political history, which is described at the end of this chapter (Section 3.14). Except for this brief period, he held his position as Extraordinary Professor until it was upgraded to that of Ordinary (Full) Professor in 1961, a year before his formal retirement.

It was a mean career, blighted by marginalization and outright humiliation, and in some ways incommensurate with the magnitude of his contributions to science. As we shall see, some of his difficulties in gaining recognition arguably were self-inflicted, but others were not. To unravel all their causes would take more research than has been done, but there is little doubt that the personal frustrations of Hückel's career mirrored the decades-long delay in general acceptance of the significance of his work.

3.3 The Debye-Hückel Theory of Electrolytic Solutions

As an assistant in Zürich, Hückel collaborated on the famous Debye-Hückel theory of strong electrolytes (1923)[5]. The details of the theory need not detain us, but even a rough outline of the approach used can convey appreciation of two major characteristics of Hückel's later independent work, particularly in the field of organic quantum chemistry: first, the identification of the significant questions for which no satisfactory answers were yet available, and second the design of a theory based upon bold simplifying assumptions, which although perhaps not rigorously justifiable at the time, nevertheless showed the way to plausible explanations of known facts and to testable predictions. The brief presentation here relies heavily on more authoritative sources.[2,6,7]

At that time, experiment had established that the activity coefficient of an ion in solution decreased with increasing concentration. Debye and Hückel conceived that this was caused by an "ionic atmosphere," in which the neighboring ions of opposite charge lowered the free energy of the ion. Similarly, this hypothesis accounted qualitatively for the observed reduced mobility of the ions when the concentration was raised. As is described elsewhere,[2,6] the first part of the theory stood up well; the second, which concerned the ionic mobilities, was subsequently shown by Onsager to have a serious, but correctible, flaw.

In the ionic mobility case, two consequences of the ion atmosphere were proposed: an "electrophoretic effect" in which the counter-ions in the atmosphere pull the solvent in the wrong direction, making it necessary for the central ion to "swim upstream," and a "relaxation effect" in which the ion is held back by the net attraction of the atmosphere itself. Unfortunately, the calculated molar conductivities did not fit the experimental data. Onsager, while agreeing with the major idea of an ion atmosphere, found that one could not ignore the difficulty with the conductivities. In his own words:

» *The relaxation effect ought to reduce the mobilities of anion and cation in equal proportion. Much to my surprise, the results of Debye and Hückel did not satisfy that relation, nor the requirement that wherever an ion of type A is 10 Å west of B, there is a B 10 Å east of that A. Clearly something essential had been left out in the derivation of such unsymmetrical results.* «[2,7]

Onsager deduced that the problem was related to the Debye-Hückel assumption of uniform straight line motion of the central ion. Once this requirement was relaxed, and the ion was permitted Brownian motion, like its neighbors, the physically needed symmetry was restored, and the calculated and experimental mobility values came into good agreement.

I want to stress that, aside from the original "unphysical" asymmetry eventually corrected by Onsager, the Debye-Hückel theory rested on a separate simplification for which no independent evidence was then available, namely that the ionic atmosphere is a static entity, in which density fluctuations could be ignored. Only later, in 1925, did Fowler justify this key assumption for very dilute solutions on statistical mechanical grounds.[2] What is really operating here in this early phase of Hückel's career is a particular style of theory in which the goal is not a perfect, unshakable construct that will last for eternity, but rather a more heuristic procedure, which might be described with the motto: *let's see if this works, and if it does, let's keep using it until it shows deficiencies.* Of course, this is the way many theoreticians often operate, whether or not they admit as much.

It is true, as has been said,[2] that we don't know for sure whether Debye or Hückel was the dominant partner in designing the approach to their joint problem, but this issue is almost beside the point here. One way or another, Hückel learned (or invented, or re-invented) this style in the work on the theory of electrolytes. It was to be a hallmark of his later independent research.

3.4 The Nature of the Double Bond

In searching for a pattern of influence on Hückel's famous 1930 papers[8] "Zur Quantentheorie der Doppelbindung," we note that he acknowledges the stimulus given by Bohr, in whose institute in Copenhagen he began the work in the summer of 1929. The first paper was finished in Leipzig at the end of the year. Hückel's autobiography[1] implies that Bohr's role consisted largely of identifying chemistry as a field to which the new quantum ideas might be fruitfully applied. There is no indication that Bohr made detailed recommendations as to the areas of chemistry that might be interesting to examine, but apparently he did suggest that understanding the nature of the double bond would make an interesting subject.

Where then did Hückel get the idea to work on the theory of such relatively sophisticated topics (for the time) as restricted rotation in unsaturated compounds and the source of "aromatic" character, the special stability of certain cyclic compounds? One might imagine that this much insight into the details of so foreign a discipline would have been unusual for a theoretical physicist of the time. It is conceivable that as a student or subsequently in preparation for his theoretical work, he may have acquired enough knowledge of chemistry, especially of organic chemistry, to generate the required motivation, but he makes no mention of special study. One can hardly avoid the conjecture that Walter Hückel's encyclopedic knowledge of the subject offered a far more accessible source. According to Erich Hückel's acknowledgment in the 1930 paper, Edward Teller was especially helpful on quantum mechanical questions and Walter Hückel on chemical matters. Our understanding of the human factors conditioning the interaction of theory and experiment would be greatly enhanced if we knew more about Walter's role. Although Erich's autobiography and other writings so far available to me are quite non-specific about just what Walter provided, it is my (as yet undocumented) working hypothesis that Walter served as more than a mere source of factual information. It seems likely that his professional immersion in the culture of organic chemistry made it natural for him to point out to Erich several profound unsolved problems of the structure and reactions of organic compounds which one could hope to illuminate with the new quantum theory. In other words, Walter was in a position to perform a crucial service to Erich: he could ask the right questions. Whether he did so remains to be established.

In his first paper on organic quantum chemistry, Erich Hückel undertook to solve an intimidatingly deep problem, which is concisely stated in his own words:[8]

» *The chemist, especially the organic chemist, tends to link more to the concept of valence than merely the valence of the atoms. He would like to ascribe to the valence lines between bonded atoms a definite real existence, in which, especially in the chemistry of carbon, not only the number of valence bonds, but also their direction in space should have significance ... In this work the generality of this question will not be treated; rather only a special case will be examined, which makes a contribution to this question. This case concerns what chemists call the 'restricted rotation of double bonds'.* «

The persistence of stereochemical configuration about C=C and C=N bonds had long been known and had been rationalized by the postulate of restricted rotation in such compounds. J. H. van't Hoff[9] had provided a classical (i. e., pre-quantum theoretical) "explanation." The carbon-carbon double bond was imagined to be made up by contact of two tetrahedrally disposed valences of each atom (Figure 2). This would result in the stereochemistry shown, with the four remaining valences lying in the same plane as the carbon atoms.

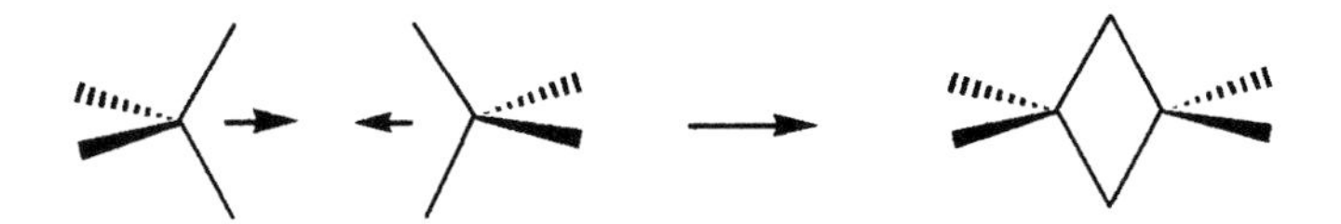

Figure 2. Junction of two sets of tetrahedral carbon valences to produce a double bond, in the manner of van't Hoff.[9] The bonds lie in the paper plane. The remaining four valences lie outside the paper plane.

The structure would be resistant to rotation of the carbons and their valences with respect to each other about the line joining them, because such twisting would diminish the contact of the valences. Of course, this proposal begged the question that Hückel was concerned with: what is the physical nature of these valences?

Soon after the first[8] of Hückel's papers on the double bond appeared, Pauling[10] and Slater[11] independently were to develop a quantum mechanical description of ethylene which was very close in spirit to van't Hoff's representation. It visualized the two carbon-carbon bonds as *equivalent entities* made up by overlap of *hybrid* sp^3 orbitals whose axes lay on either side of the actual C–C bond direction. As part of a general theory of directed valence, these same sp^3 hybrid orbitals had been postulated to account for the tetrahedral orientation of the bonds of tetravalent carbon, and hence for much of the stereochemistry of aliphatic compounds.

Several textbook authors have wrongly ascribed to Pauling the concept of trigonal (sp^2) hybridization of the doubly bonded carbons in ethylene (see below). Pauling's example of this hybridization was graphite, not ethylene. In fact, he remained strongly opposed to the concept that the two ethylene C–C bonds were different, an unavoidable consequence of trigonal hybridization. The angle H–C=C in a trigonally hybridized alkene is predicted to be 120°, but Pauling's *The Nature of the Chemical Bond* (1961 edition),[10] citing a number of experimental determinations near 125° 17′, stoutly defends the tetrahedral model, which predicts the latter value (half the difference between 360° and the tetrahedral angle H–C–H angle 109° 26′). It is probably fair to say that in the current era of ab initio theory, a decision between these two approaches, which are after all only *models*, has become moot.

Curiously, one often sees, in implied or direct form, similar misattributions of the trigonal hybridization concept to *Hückel*. It is true that the essence of Hückel's model is that the two C–C bonds are *nonequivalent*, one σ and one π, a basic distinction from the tetrahedral model favored by Pauling. However, Hückel's ethylene model in the first instance did not arise from hybridization considerations and did not require trigonal hybridization. In fact, the σ–π ethylene model originated in the quantum theoretical treatment of a seemingly very different molecule, molecular oxygen, O_2. How Hückel's new plant sprang from such unlikely soil makes a long story, but the lessons to be learned about the genesis of ideas justify an abbreviated version.

In 1929, the year just preceding the appearance of Hückel's paper, Lennard-Jones[12] had made a molecular orbital analysis of the electronic structure of dioxygen in its ground state, which he represented as having the occupation pattern in (1):

$$(1s)^2\ (1s)^2\ (2s)^2\ (2s)^2\ (2p_+)^2\ (2p_-)^2\ (2p\sigma)^2\ \{2p\pi_+,\ 2p\pi_-\} \quad (1)$$

As Hückel pointed out,[8] this notation differs from the "united atom" MO formalism of Hund and Mulliken, which is more familiar now, and which had been developed for diatomic molecules. Lennard-Jones's notation is more suitable for showing the electronic states that are being generated during dissociation of the "united atom." The results are shown here in Figure 3.[13]

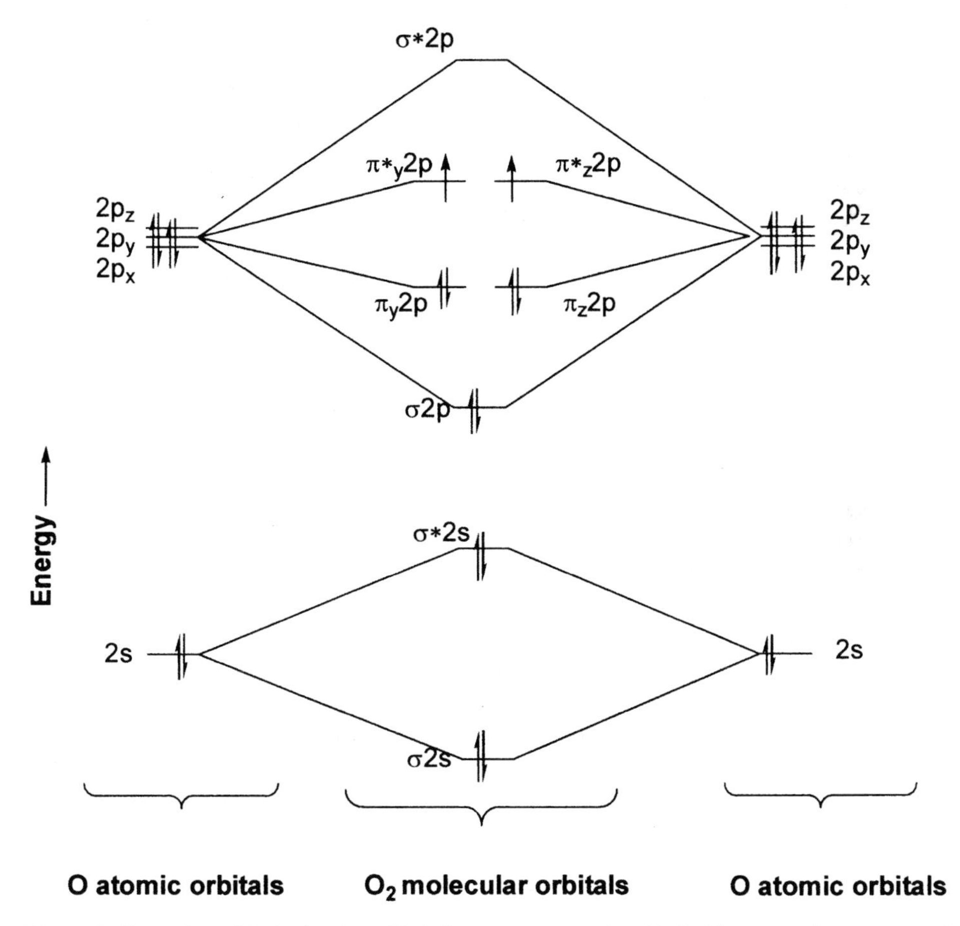

Figure 3. Formation of O_2 molecular orbitals from oxygen atomic orbitals. The occupation pattern and spin state are derived by application of Hund's highest multiplicity rule. The energies are schematic. Adapted from C. A. Coulson, *Valence*, Oxford University Press, Oxford, 1952, p. 100, with permission of Oxford University Press.

Note that the interaction of the two filled 2s atomic orbitals gives rise to no net bonding in this approximation, which (with apologies) limits the discussion to the molecular orbitals (MOs) derived from the atomic 2p-orbitals and alters Lennard-Jones's presentation to a "united atom" form. An oxygen atom in its ground state has four electrons in the three mutually perpendicular atomic p-orbitals. When two such atoms are brought into bonding distance, the six atomic orbitals give rise to six MOs. Two of the MOs, a strongly bonding (σ2p) and a strongly anti-bonding (σ^*2p) pair, result from mixing of the two original p-orbitals that lie along the internuclear line. The remaining four atomic p-orbitals mix to form two degenerate π-MOs and two degenerate π^*-MOs. The nodal planes of the π-MOs are a mutually perpendicular (arbitrarily chosen) pair containing the two atomic nuclei, and the π^*-MOs each have the same pair of nodal planes in addition to a perpendicular nodal plane bisecting the internuclear O–O line. The total of eight (originally p) electrons now must be fed into this set of MOs. Two each, with opposed spins, go into the bonding σ and the two bonding π MOs. Although formally the remaining two electrons can go into the pair of π^* orbitals in any of several occupancy and spin configurations, Lennard-Jones assumed that, in energy, "that state is held to be lowest which has the greatest multiplicity, as is the case in atoms." Here, this would be a triplet state, in which the last two electrons each occupy one of the degenerate π^*_y and π^*_z orbitals and their spin vectors are parallel. The nominal overall bond order is two, since although there are three bonds, one of them can be considered to be formally cancelled by an anti-bond.

By this explanation of the experimentally known fact that dioxygen is paramagnetic in its ground state, Lennard-Jones's analysis provided a brilliant early triumph for the quantum theory. Apparently, Hund's multiplicity rule,[14,15] which originally had been promulgated for atoms, also applies to certain molecules. Dioxygen is a case in which the rule might be expected to apply strictly, because of the symmetry-enforced degeneracy of the π^*-orbitals (see below). Nevertheless, this insight was not the major focus of Hückel's interest in Lennard-Jones's paper. In fact, the triplet nature of dioxygen might well have bemused a lesser intellect, and a connection between the intriguing special case of dioxygen and the less spectacular but far more general problem of ethylene might never have become clear. What attracted Hückel was the idea that there can be *two kinds* of oxygen-oxygen bonds, σ and π.

The next step toward the description of doubly bonded carbon was an ingenious *gedanken* experiment. Hückel imagined the conversion of one of the oxygen nuclei of dioxygen to carbon by the extraction of two protons, which then were bound to the resulting carbon nucleus to give formaldehyde. If we today were following this protocol, we probably would make formaldehyde analogously to the construction of dioxygen shown in Fig. 3, with the exception that one of the oxygen atoms would be replaced by an sp^2-hybridized fragment CH_2, and the units would be brought together so that all four atoms were in a common plane. Again, one of the carbon 2p-orbitals could form a σ-bond by overlapping a 2p-orbital of the oxygen, and the other carbon 2p-orbital would form a π-bond with the oxygen p-orbital. However, Hückel declined to make the assumption of coplanarity at the start, and argued his way through alternative geometries before rejecting them in favor of the planar one.

Also, in 1930, it was not obvious what should be the ordering of the energies of the MOs resulting from the procedure. Hückel again presented arguments that favored the ordering we accept today.

The details, though interesting, are too lengthy to give here, but one major point is worth emphasizing. In both arguments, Hückel resorted to *experimental* data to make his decisions. With regard to the multiplicity question, it was clear that because of the need to bind the substituent hydrogens to carbon in formaldehyde, the π^* MOs would not be each singly occupied, and the degeneracy that causes a triplet ground state in dioxygen would not exist in formaldehyde. Nevertheless, it was (then) a difficult computational problem to decide whether the multiplicity of the ground state of formaldehyde should be a singlet π or a triplet π–π^*. Hückel's basis for rejecting the triplet was analogical: although no experimental information on the magnetic properties of formaldehyde was available, it was known that its close relative acetaldehyde was diamagnetic and hence a singlet.

Similarly, with regard to the molecular geometry, note that a configurationally stable pyramidal monosubstituted formaldehyde, *e. g.*, acetaldehyde, with three non-equivalent substituents, would predict the existence of optical isomers. No such isomers were known, as Hückel pointed out. It is heartening to realize that Hückel, a pioneer of theory, was in the latter case not too proud to use the organic chemist's traditional (but non-rigorous) argument of isomer numbers! We have no evidence that Walter Hückel provided this insight to his brother, but it is easy to imagine his doing so.

Of course, the choice among theoretical possibilities by appeal to experimental fact had ample precedent in quantum mechanics. A famous case[16] is the postulate by Heisenberg and Dirac, based upon experimental spectroscopy, that "only the antisymmetric solution [of the total wave function] existed in nature and that the symmetric solution represented a solution which we do not find in the ordinary universe."[16] It was this assumption from experiment, rather than a theoretical derivation, that provided the basic explanation for the Pauli exclusion principle.

Formaldehyde-to-ethylene then required only a repetition of the oxygen-to-CH_2 *gedanken* step. Again, the presence of the substituent hydrogens is the key structural feature that favors a singlet ground state. Hückel had now achieved his first objective, a quantum mechanical description of the double bond that would explain restricted rotation. In his model, the C=C double bond is made up of a σ and a π bond. The σ bond is axially symmetric about the internuclear C–C line, but the π bond is not. Rotation of one of the CH_2 groups out of the plane of the other is resisted because to continue this process to 90° would require breaking a π-bond. Hückel did not provide a quantitative estimate of the strength of this bond, but it must be substantial to account for the thermal stability of olefinic cis-trans isomers. He emphasized the deep structural difference between this model and that of van't Hoff, in which both of the C–C bonds are equivalent.

3.5 Hybridization in Double Bonds

It will be clear that a bit of mystery still lingers over Hückel's picture of ethylene at this point: What is the nature of the C–H bonds? Hückel never clarified this point but left it to others. In 1931–1932, the years immediately following Hückel's first paper on ethylene, the concept of quantum mechanical hybridization was introduced as the basis of a theory of directed bonding. As we have seen, this idea played an important role in the development of the Pauling-Slater equivalent-bond model of ethylene. In 1933, Mulliken[17] suggested trigonal hybridization of the carbon valences in ethylene, and in 1934, Penney[18] made a more extensive comparison of this model with two alternatives, the so-called "right-angle" model and the van't Hoff-like Pauling-Slater tetrahedral model.

In the "right angle" model (see Figure 4), Penney imagined the C–H bonds to be pure 2pσ, and the double bond joining the two carbons to be made up of one (s, s) and one (σ, σ) bond. The vicinal hydrogen interactions are neglected, and the C–C bonds, although not equivalent, are both axially symmetrical about the C–C direction, so that the energy of the configuration does not depend on the the angle ϕ through which one carbon is rotated with respect to the other. As Penney recognized, this model predicts free rotation about the C=C bond and therefore cannot account for one of the characteristic properties of alkenes.

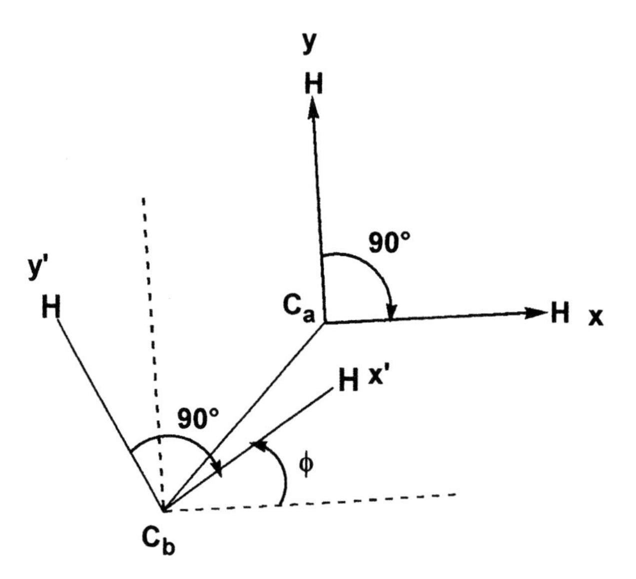

Figure 4. Penney's "right angle" model (ultimately rejected in favor of the σ-π model) for ethylene. The C–H bonds are pure 2pσ, and the double bond joining the two carbons is made up of one (s, s) and one (σ, σ) bond. The vicinal hydrogen interactions are neglected, and the C–C bonds are axially symmetrical about the C–C direction, so that the energy of the configuration does not depend on the the angle ϕ through which one carbon is rotated with respect to the other. Adapted from ref. 18a by permission of the Royal Society.

Penney compared the tetrahedral model and the trigonal model by means of a valence bond calculation and concluded that the trigonal model was energetically preferred. Although the sp^3 hybrid orbitals of the tetrahedral model have more extension along the orbital axis than the sp^2 orbitals of the trigonal model, the sp^3 orbitals are necessarily canted outward from the internuclear C–C line and hence overlap poorly. He subsequently proposed trigonally hybridized carbon as the building block of the σ-framework of the benzene ring.

Eventually, these suggestions became widely adopted for discussions of doubly bonded carbon. In part, this acceptance must have been furthered by later group theoretical arguments by Kimball,[19] but one may speculate that it also was based on the simple conceptual continuity of the hybridization picture (sp^3 for methane, sp^2 for ethylene, sp for acetylene), which made the idea pedagogically attractive. Actually, a more hard-eyed evaluation of the quantitative merits of Penney's calculations might have been appropriate. One certainly should have been concerned about Penney's conclusion that the energy needed to rotate one CH_2 plane in sp^2 hybridized ethylene by 90° with respect to the other was "quite small, probably about 1/2 volt,"(about 11.5 kcal/mol). If a similar value were required in substituted ethylenes, stable cis-trans isomerism could not have been observed at room temperature. In other words, were one to take the calculation at face value, the same argument used to reject the "right angle" model logically would have required rejection of the trigonal model.

3.6 The Benzene Problem

It might be surmised that Hückel proceeded from ethylene to conjugated chain compounds such as allyl, butadiene, pentadienyl, hexatriene, etc., and then to benzene and other cyclopolyenes, in the systematic manner that we teach Hückel molecular orbital (HMO) theory to students today. Actually, this is not the sequence that occurred. Hückel eventually considered unsaturated chains as a class,[20] but in the breakthrough paper in 1931[21], Hückel's target was benzene itself.

This molecule had long been recognized as the parent substance of a series of compounds which had in common certain properties called "aromaticity". The precise meaning of the term "aromaticity" has tended to evolve over the years, but among the chief properties attributed by chemists to the aromatic compounds were special chemical stability and persistence compared to models derived from the properties of simple alkenes or cycloalkenes. In the late nineteenth century, the association of this property with *cyclic conjugation* was recognized most clearly by Thiele (see Chapter 2), and in 1925, Armit and Robinson[22a] put forward the idea that it was specifically the closed loop of *six* electrons in the conjugated system, an "aromatic sextet," that conferred special stability upon the benzenoids. Since then, the definition of aromaticity has continued to evolve, and there has even been some discussion of whether there is such a thing as aromaticity.[22b] One is tempted to draw an analogy to the case of pornography, of which Supreme Court Justice Potter Stewart is said to have remarked that he could not define it but he knew it when he saw it.

Hückel's first paper[21] in this field was his *Habilitationsschrift* for attaining the *venia legendi* (right to teach) in theoretical physics at the Technische Hochschule Stuttgart. A massive document of 84 printed pages, it gave two descriptions of benzene and other conjugated cyclopolyenes: the so-called "first method," which eventually came to be the valence bond (VB) theory, and the "second method," the application of the molecular orbital (MO) theory. Hückel gave reasons for preferring the MO procedure, and although some of these might not be very convincing today, he persisted in this choice thereafter. Both the "first method" and the "second method" assumed that the unique properties of cyclic conjugated systems could be attributed approximately to the π-electrons without explicit consideration of their interaction with the σ-electrons.

3.6.1 Hückel's "First Method"

The VB method, which Hückel had rejected, was taken up soon after by Pauling and others,[10,23–25] who simplified the mathematical procedures and gave reasons for preferring the VB to the MO method. Again, with regard to the problem of cyclic conjugated molecules, their reasons seem insufficient today, but these workers held unwaveringly to their choice in later years. This "classical" VB method, and Pauling's pedagogical packaging of its major results in the form of the "theory of resonance,"dominated theoretical chemistry for 25 years after 1930. The reasons for this, further examined below, were not necessarily that the theory was "correct" in any absolute sense, but rather that Pauling, with consummate knowledge of the whole field of chemistry, convincingly showed how a broad range of chemical phenomena, especially in small molecules, could be explained by *some kind* of quantum ideas[10] and thereby established a faithful and largely uncritical following. As it happened, however, the classical VB method, in the truncated form of it he used, was to prove theoretically inadequate when applied to the problem of aromaticity.

It is true that with the emergence of high-speed digital computers, as is described elsewhere,[26,27] the MO method, in increasingly sophisticated manifestations, gradually became the major basis for the explosive computational development of electronic structure theory. However, one should not conclude that VB theory is without adherents today. On the contrary, many investigators have contributed developments of advanced VB methods and successful applications to significant problems, including as we shall see (Section 3.6.2), the reconciliation of the VB and MO theories of aromaticity.[27–29]

The most characteristic feature of the valence bond method is that it considers the combining atoms as a whole.[30] The formation of a molecule is thought of as arising from the bringing together of complete atoms, which are then allowed to interact. In this it differs from the MO method, in which only the nuclei (or nuclei + electron inner shells) are first brought into position, and afterwards the valence electrons are allotted to polycentric molecular orbits. Clearly, the VB method corresponds more closely with the conventional chemical picture, which probably accounts for its widespread acceptance in the form presented by Pauling.

Hückel's "first method," the early form of VB theory, was derived from Heitler-London[31] theory for the formation of molecular hydrogen from two hydrogen atoms. It starts with each atom in a specified quantum state and then introduces exchange. The exchange procedure, which takes into account the indistinguishability of the two electrons 1 and 2 in the form of the exchange integral $<\Psi_A (1) \Psi_B (2) | H | \Psi_B (1) \Psi_A (2)>$, has major significance in the VB method. For example, the binding energy of hydrogen calculated without exchange is only 5.5 kcal/mol but when exchange is included in the wave function, the binding energy rises to 69 kcal/mol, a substantial fraction of the experimental value of 104 kcal/mol.[30]

By application of group theoretical procedures, Heisenberg[32a] used this exchange method to explain ferromagnetism. Slater[32b] then developed a useful determinantal method which made possible the treatment of interaction among a large number of atoms without group theory and applied it to electrons in metallic lattices. This method was further applied by Bloch[33] in place of the Heisenberg group theoretical method for the theory of ferromagnetism. Hückel used the Bloch formalism directly in his "first method."

As described by Hückel,[34] the treatment of conjugated molecules by the classical VB method begins by assigning to each carbon atom a π-electron which is in a given state with the positional eigenfunction $\phi_a (r_{ia})$, where the suffixes indicate the *i*th electron in the atom *a*. The total positional eigenfunction, taking coupling of the π-electrons into account, is written as a linear combination of the products $\phi_1 (r_{i1}) \ldots \phi_6 (r_{i6})$. Starting from this, the Heitler-London perturbation method is worked out to the first approximation. Spin is taken into account, and only the linear combinations that conform to the Pauli principle are considered.

In the Pauling modification of VB theory,[23–25] one selects from these linear combinations those which correspond to the smallest value of the total spin (in this case $S = 0$), and chooses from these the ones that are linearly independent. These functions which belong to the value $S = 0$ can be associated with models of the valency pattern of the π-electrons in which the atoms are joined in pairs by single bonds, one and only one bond radiating from each atom in such a way that the bonds do not cross one another. The schemes of valencies corresponding to the functions Pauling called "canonical structures." One then formulates and solves the secular perturbations corresponding to these functions, restricting the coupling to that between *adjacent* atoms.

Wheland[23d] gives a simple example of the application of this method to the case of cyclobutadiene. The canonical structures are **A** and **B** (Figure 5).

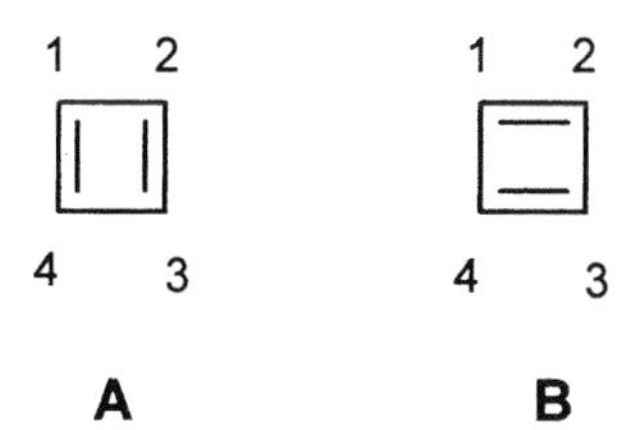

Figure 5. Canonical structures of cyclobutadiene

One designates the eigenvalue of the energy as W, the 2p function on the *i*th carbon atom as Ψ_i, the coulomb integral $<\Psi_1\Psi_2\Psi_3\Psi_4 \mid H \mid \Psi_1\Psi_2\Psi_3\Psi_4>$ as Q, and the single exchange integral between adjacent carbons, for example, $<\Psi_1\Psi_2\Psi_3\Psi_4 \mid H \mid \Psi_2\Psi_1\Psi_3\Psi_4>$ as α'. All single interchange integrals of the energy between non-adjacent atoms, all multiple interchange integrals, and all interchange integrals of unity are neglected. The Slater valence-bond eigenfunction may be expressed as a secular equation (2):

$$\begin{vmatrix} Q + \alpha' - W & (1/2)\,Q + 2\alpha' - (1/2)\,W \\ (1/2)\,Q + 2\alpha' - (1/2)\,W & Q + \alpha' - W \end{vmatrix} = 0 \qquad (2)$$

This has the solutions $W = Q + 2\alpha'$ and $W = Q - 2\alpha'$, of which the former represents the ground state since the exchange integral α' is presumably negative. The resonance energy is obtained by subtracting from this a quantity $Q + \alpha'$,the energy of a single one of the two canonical structures: $Q + 2\alpha' - (Q + \alpha') = \alpha'$.

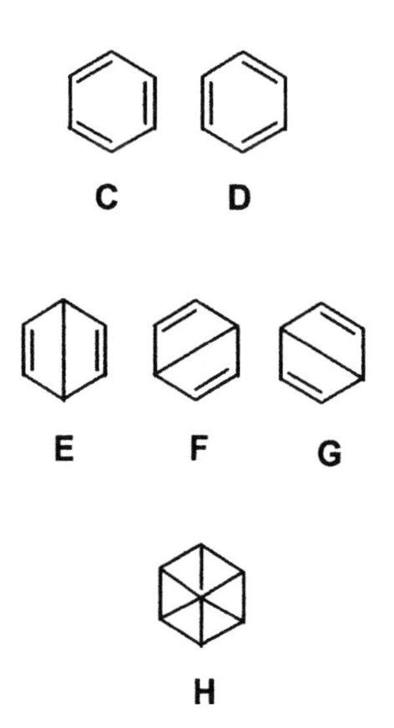

Figure 6. Canonical structures of benzene.

In the case of benzene, there are five canonical structures (Figure 6) with corresponding canonical functions: **C** and **D** correspond to the Kekulé forms, **E, F** and **G** to the Dewar forms. Other structures, such as for example **H**, that of Claus, can be expressed as linear combinations of the canonical functions (3):

$$\phi_H = \phi_C + \phi_D - [\phi_E + \phi_F + \phi_G] \qquad (3)$$

Following a procedure similar to that used for cyclobutadiene, the coupling energy for benzene is $6Q + 2.6055\alpha'$, and the molecular eigenfunction is (4).

$$\Phi_{\text{benzene}} = 0.62435\ (\phi_C + \phi_D) + 0.27101\ (\phi_E + \phi_F + \phi_G) \tag{4}$$

In the sense of the approximation, the ground state thus may be considered to result from the superposition of the two Kekulé and the three Dewar forms (Figure 6).

One might be concerned about the approximation which limits the exchanges to pairwise nearest-neighbors. Clearly, there are many more permutations that could be included. Wheland[23d] himself states:

» *This assumption is an extremely drastic one and no rigorous justification for it can be given. The integrals we ignore are probably, to be sure, rather smaller in magnitude than the other ones which we retain. There are, however, an enormous number of the former integrals, and, even though some of them are positive and some are negative, we can have no assurance that their net effect is negligible. We shall, nevertheless, make the assumption because, without it, our calculation would become impracticably complicated, and because, with it, surprisingly satisfactory results can be obtained.* «

As we shall see, the neglect of some of these other permutations has serious consequences in the description of cyclic conjugated systems.

Although the classical VB procedure came to be called[23d] the "Heitler-London-Slater-Pauling (HLSP) method," Slater in fact preferred to put some distance between himself and Pauling in applications of the method. Thus, in a discussion[35] at the International Conference of Physics in 1934, Slater remarked that he was in "entire agreement" with Hückel and Hund that the MO method is superior to the Heitler-London method for computing interatomic forces, and that "the calculations of Pauling, for instance, seem to make quite unwarranted use of the theory." Slater gave no further details, but it seems a reasonable conjecture that he was expressing misgivings over the truncation.

3.6.2 Hückel's "Second Method"

This procedure, subsequently called[23d] the "Hund-Mulliken-Hückel (HMH) method," had as an immediate intellectual precursor another study[36] by Bloch on the quantum mechanics of electrons in crystal lattices. Bloch, at that time in Leipzig, was actively developing the theory of metals in order to understand such phenomena as conduction and magnetism. Subsequently, of course, he became famous for providing some of the theoretical foundations for the field of nuclear magnetic resonance.

Bloch approached the problem of the electronic interactions in a many-electron system by considering the properties of a single electron in a spatial force-field perturbed by the atomic nuclei and the statistical charge distribution of the remaining electrons. The idea is very similar to the Hartree-Fock method for treating many-electron atoms. In Bloch's sinewy words, "the force-field has the same periodicity as the crystal lattice itself," and "we are thus dealing with plane de Broglie waves which are modulated in phase with the lattice."

The procedure Hückel used[20] was to set up such a lattice for benzene, calculate the MOs (wave functions) and energy levels in terms of the parameters α and β (defined below) by solving the Schrödinger equation, and construct the electronic configuration by *aufbau*, in accord with the Pauli principle and in analogy to the practice of Hund, Mulliken, and Lennard-Jones for diatomic molecules. Hückel's benzene lattice was a regular hexagonal array of carbon $2p_z$ orbitals. As every chemistry student now knows, the results can be expressed as shown in Figure 7, where α is the energy of an electron in an unperturbed carbon 2p orbital, and β is the stabilization energy (relative to α) experienced by an electron when two such units interact at a defined distance.

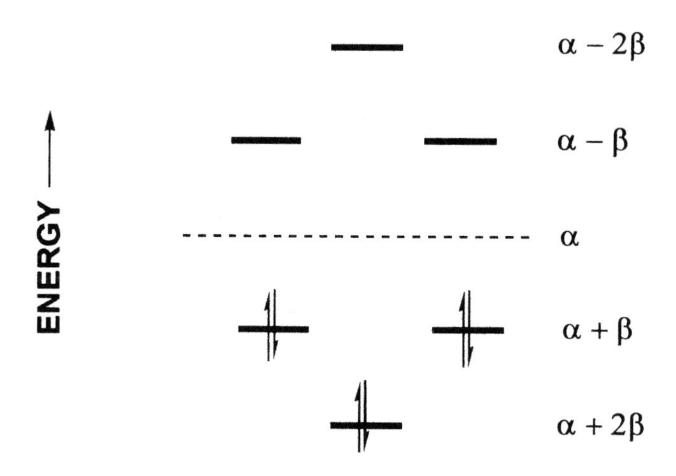

Figure 7. Energy levels of benzene and occupation pattern in the ground state as calculated by Hückel's second method. The parameters α and β are defined in the text.

Hückel generalized this result to other conjugated systems with n centers, both to rings of other sizes and to chains.[20,21,34,37] The energy levels can be expressed in the famous Hückel equations (5) and (6), where $m = (n-1)/2$ for n odd and $n/2$ for n even:

$$\textbf{Chains } C_nH_{n+2}\text{:} \quad E = \alpha + 2\beta \cos \frac{j\pi}{n+1}, j = 1,2,3..n \tag{5}$$

$$\textbf{Rings } C_nH_n\text{:} \quad E = \alpha + 2\beta \cos \frac{2k\pi}{n}, k = 0,\pm 1,\pm 2,..m \tag{6}$$

Among the messages that these equations convey are that the energy levels for the chains are unique, whereas some of those for the rings are two-fold degenerate, that is, the eigenvalues sometimes occur in pairs of equivalent energy.

As Hückel pointed out,[37] these degeneracies are a direct consequence of the cyclic nature of the electron circulation in the ring compounds. The physical reason is that in the energy states which have a finite angular momentum associated with the electronic motion, the circulation is *directional* in either a clockwise or a counterclockwise sense. The eigenfunctions corresponding to these states are not identical, even though the energies are, so *two* MOs must exist at that energy. An equally important point is that for n even or odd the lowest eigenvalues of the ring compounds are unique, and for *n* even, the highest also are unique. These are non-intuitive results which come out of the solution of the Schrödinger equation and which mark this approach as fundamentally different from the old (Bohr) quantum theory. Whereas in the old theory, for example, the lowest atomic s-states still had angular momentum $h/2\pi$, in the new theory the s-states now have zero angular momentum. Similarly, in the unique states of benzene, the angular momentum is zero, and consequently, the "sense" of the electronic circulation becomes meaningless.

This difference between rings and chains will be familiar to students of elementary quantum mechanics,[38–41] where a favorite pedagogical exercise is the calculation of the eigenvalues of the so-called "one-dimensional" Schrödinger equation. An electron is imagined to be constrained to move along a line or in a circle under zero potential. The solutions for the linear case are unique; those for the circular case have the lowest energy level unique but all the rest pair-wise degenerate.

Hückel[20,21,34,37] emphasized that the MO method, in producing such energy level patterns, led to the idea of closed electron shells analogous to those of the noble gas closed shells in atoms. Particularly significant for the *n* even conjugated rings was the case of six electrons in benzene, which had been recognized for some time as a feature leading to aromatic properties. Since the Pauli principle allows each orbital to accommodate two electrons of opposite spin, filling the first three orbitals (first two energy levels) of the benzene set 1, 2, 2, 1 in this way would give a stable configuration. On the other hand, four electrons in cyclobutadiene would not give a closed shell, since only the lowest level of the 1, 2, 1 set could be filled. The next higher level could accommodate four electrons but only would have two available. Morover, in the case of ions derived from odd-membered rings, with all levels above the lowest being degenerate, a closed shell configuration again is reached with six electrons, as in cyclopentadienide anion, but not with eight, as in cycloheptatrienide anion or with four as in cyclopentadienium cation.

This rule, subsequently stated by others in the abbreviated form $(4N + 2)$ $(N = 0, 1, 2\ldots)$, namely that aromaticity should be associated with monocyclic π-electron systems containing 2, 6, 10... electrons, was in accord with the known facts of organic chemistry at the time. In addition, predictions now could be made of the existence and properties of unknown but easily imaginable new structures. Although the rule does not apply strictly to polycyclic compounds such as naphthalene or biphenyl, Hückel showed by explicit MO calculations[37] that these systems too should be considered aromatic, as might have been expected. There are limitations to the rule, but much computational research in recent years confirms the essential fact that Hückel's broad classification of aromatic and antiaromatic character survives at higher levels of MO theory and is not just an artifact of the approximations used in the early treatment.[42]

Significantly, the classical VB method did not produce these results. For example, it predicted that cyclobutadiene should have the highest resonance energy per electron of any of the even cyclic polyenes, and it gave no reason to expect that cyclopentadiene should be a much stronger acid than cycloheptatriene. These failures of classical VB theory were among the reasons that Hückel in 1931 turned away from it as a basis for understanding aromaticity. More than half a century later, the opinion persists[43,44] that classical VB theory is "decisively unsuccessful"[44] for this purpose.

Nevertheless, one feels intuitively that a higher level of VB theory should be capable of reproducing the Hückel rule. After all, both the MO and the VB approaches are approximations of the complete solution of the Schrödinger equation. At successively higher degrees of approximation, the two methods should converge onto equivalent results. In fact, such convergence can be shown analytically for a simple molecule such as H_2 and is expected generally, however complicated the molecule may be.[45] It comes as no real surprise, therefore, that eventually methods for deriving the Hückel rule by VB theory should emerge.[44,46,47] The key insight[44,46,47] in solving this problem is the crucial necessity to include *cyclic* permutations of the π-electrons in the exchange procedure. Just such exchange integrals were among those omitted in the classical VB method. An experimentalist senses a gratifying propriety in this result.

3.7 A Chilly Reception from the Experimentalists

Hückel's pioneering papers on the molecular orbital theory of unsaturated and aromatic compounds appeared in the period 1930–1931, but they seemed to make little impact on the community of chemists for many years after. In their biographical memoir of Hückel, Hartmann and Longuet-Higgins[2] ascribe the general neglect of his results to the national culture of German science:

» *... physicists in that country in any case were not ready to accept investigations about more complicated chemical bond phenomena as a typical contribution of a physicist. Still more difficult was his (Hückel's) relationship to the chemists. Before World War II, especially in the Anglo-Saxon countries, chemical physics and within that field quantum chemistry also was accepted by both physicists and chemists as an interesting new field of science. Chemists in Germany, on the other hand maintained that chemistry is what chemists do. They did not do quantum chemistry and therefore this kind of science did not belong to chemistry.* «

It may well be true that German chemists of that period or even later resisted quantum ideas. Although Rolf Huisgen reports[48] that he was teaching Hückel's results (but not the details of how the calculations were done) as early as 1949, this undoubtedly was exceptional. Even Walter Hückel makes only cursory reference in his textbook[49] to the contribution his brother had made to the problem of the aromatic sextet. However, I regret to say that crediting a more receptive attitude to "Anglo-Saxon" chemists, especially the experimental organic chemists, pays them a higher compliment than most of them deserve. Again, there were scattered exceptions. For example, Ingold[50] in Britain and Hammett[51] in the United States clearly were aware of the importance of Hückel's work. Elsewhere in these countries, however, even though the problem of aromaticity was prominent in the minds of chemists, and much experimental effort was devoted to it, whatever theoretical reasoning experimentalists brought to bear on such issues in their research or teaching depended almost entirely on resonance theory.[52] Noller's review of 1950,[53] ostensibly an exposition of molecular orbital theory prepared for the edification of organic chemistry teachers, does not cite a single reference to Hückel. Most elementary and even advanced textbooks of those years make no mention whatever of Hückel's ideas. One of the most influential books in the field, the second edition (1943) of the multi-authored *Organic Chemistry An Advanced Treatise*, edited by Gilman, contains a substantial chapter by Fieser on aromatic compounds.[54] The question of aromaticity is presented only from a historical perspective, and for a theoretical rationale, Fieser defers entirely to Pauling, who contributes a chapter on theory in the same work,[23c] using the already well known resonance method. Although Pauling mentions the Hückel theory there, he declines to discuss it further on the grounds that the resonance approach "is the more closely related to the usual concepts of chemistry." Not surprisingly, Pauling's omissions of scholarly exposition and comparison here and elsewhere infuriated Hückel.[1]

It is true that there was a flurry of theoretical activity, especially in Britain, following up further implications of some of Hückel's ideas. This included work[55] by, among others, Coulson on mobile bond orders, Coulson and Rushbrooke on alternant π-conjugated systems, Longuet-Higgins on non-Kekulé molecules, and Dewar on a range of chemical properties deducible from perturbational MO considerations. These papers, written in the elegant lapidary style familiar to mathematicians, and often published outside the conventional chemical journals, were important, but they were either unknown or largely unintelligible to many organic chemists.

Similarly, a key paper by Mulliken[56a] and an important early textbook by the Pullmans,[56b] both in French, influenced a few young theoreticlly able workers such as Simonetta, and through him, Heilbronner,[57] but their immediate impact on the thinking of most organic chemists was not great.

Having lived through that period, I can attest that the attitude of many American organic chemists toward MO theory was uninformed and indifferent, if not hostile. Few of them would have taken the trouble to slog through Hückel's highly technical papers, which bristled with equations and matrices, and which, with one obscurely placed exception,[34] were not available in English. The charismatic Pauling had provided them with a theory which at some level required no mathematics and could be applied using familiar bond structures. Most of them were content with that.

3.8 Experimental Tests of the MO Description of Conjugated Cyclic Compounds

If organic chemists were to a large extent unaware of the Hückel rule, how could there have been deliberate tests to explore its scope? In fact, a number of molecules synthesized or discovered in nature after Hückel's papers in the early 1930s *eventually turned out to be* relevant test species, even though the authors originally had some other motivation for studying them. In this inadvertent process, we see again the familiar paradoxical sequence encountered so frequently in science: "Here's the answer, what's the question?"

Early examples of this come from the chemistry of the azulenes {see Figure 8), a fascinating group of blue or violet substances, several of which, including the parent compound, are found in nature as such or are formed by chemical transformation of hydroazulenic sesquiterpenoid precursors.[58–60]

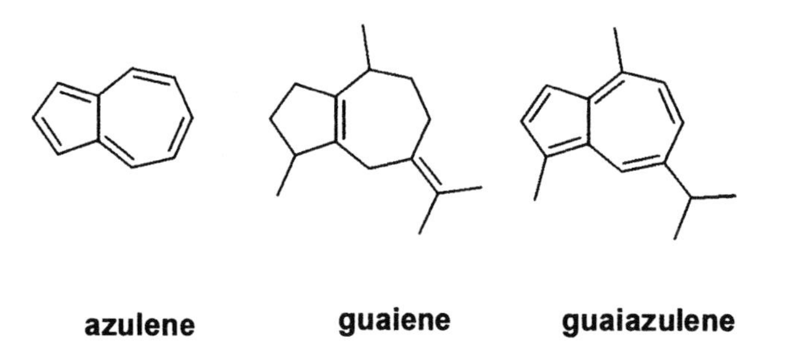

Figure 8. Typical azulenes.

The substantial variety of natural and synthetic azulenes available from the work of Plattner[58] and others[59] was augmented by the powerful new Ziegler-Hafner synthesis,[60] which made azulenes accessible in a small number of steps without the difficult final dehydrogenation previously employed. A rich store of facts embodying the chemical and physical properties of these substances now called for theoretical rationalization, which, as has been instructively summarized by Heilbronner,[61] Hückel MO theory ultimately provided.

This theme of answers anticipating questions lies at the heart of one of the inspiring stories of organic chemistry, the discovery of the troponoids.[62,63] Space does not permit a full recounting here, but a brief outline may suffice to make the relevant point. In 1926, the Japanese chemist Tetsuo Nozoe settled in Formosa (now Taiwan), where he was to live and work for the next 22 years. During the period 1944–1947, he and his co-workers had deduced the structure of hinokitiol, an isopropyltropolone (Figure 9), but because of disruptions caused by the war and the remoteness of his location, he was not aware of related activity elsewhere,[63] nor did others know of his work on this subject, most of which was not published in readily accessible journals until 1950.

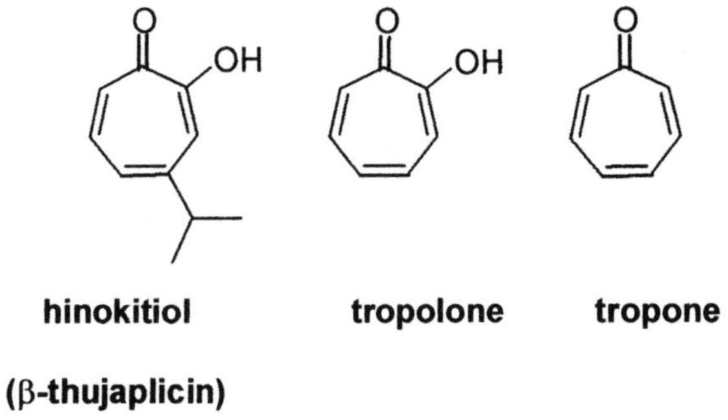

hinokitiol **tropolone** **tropone**

(β-thujaplicin)

Figure 9. Some troponoids.

Nozoe[62,63] recounts his attempts to explain by the theory of resonance the peculiar aromatic properties of hinokitiol and of tropolone itself, which he subsequently synthesized. His first exposure to this form of quantum theory came in 1942–43, when copies of the 1940 edition of Pauling's *Nature of the Chemical Bond* became available in Formosa. However, although one could write resonance structures, this gave no real enlightenment on the aromaticity of tropolone, and the matter remained somewhat mysterious to him until 1951. In that year, simultaneously published papers by Dauben and Ringold[64] and by Doering and Detert[65] described the first syntheses of tropone. Both papers called attention to the electronic structure of this ketone and emphasized the presence of a potential aromatic sextet. The Doering paper,[65] giving an explicit reference to Hückel's 1937 summary[37] of π-electron theory, pointed out that the sextet was but one of a general class of stable configurations characterized by having $(4N + 2)$ π-electrons (To my knowledge, this was the first time that Hückel's rule was stated in this succinct form). Nozoe reports[62] that at that time, he had not heard of such a rule, but it was immediately obvious to him that tropolone's aromaticity must be derived from the same source.

Clearly, the two tropone papers were examples of directed rather than serendipitous tests of π-electron theory, as were numerous others by many authors that followed.[66] As π-electron MO theory over time became refined and adapted to machine computation, so too did experimentation become more imaginative and creative in response to the growing interaction with theory. Nevertheless, the most persuasive pieces of evidence leading to the acceptance of Hückel's ideas on aromaticity seem to have been the syntheses of tropylium[67a] and cyclopropenylium[67b] cations

3.9 Orbital Symmetry, the Extension of Cyclic π-Electron MO Theory to Transition States of Pericyclic Reactions[68,69]

Periyclic reactions, as the name implies, are those "in which all first-order changes in bonding relationships take place in concert on a closed curve."[66] It would be natural to conjecture that Hückel MO methods, which had been successful in rationalizing the behavior of *ground state* conjugated cyclic molecules, also might give a fruitful account of the *transition states* of pericyclic reactions.

The idea of applying π-electron theory to reactions with cyclic transition states probably was first put forth by M. G. Evans in 1939[70]. He treated the six-electron transition state of the Diels-Alder reaction as an analog of benzene and actually wrote down a secular determinant for it differing from that of benzene only in the 1–6 and 4–5 bond integrals. In his words,

» *... very qualitatively we may say that whereas in the initial state the mobile electrons are those characteristic of an ethylene and a butadiene structure in the transition state they simulate the behaviour in a benzene structure.* « (see Figure 10).

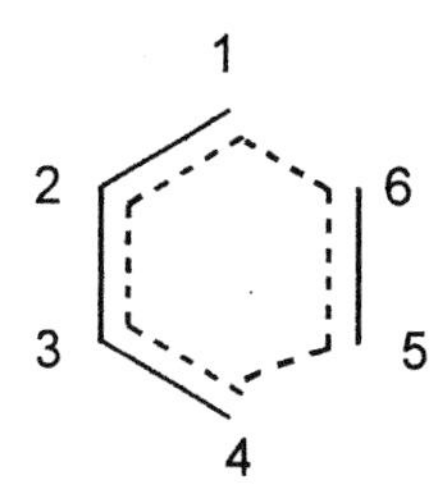

Figure 10. Evans's model of an aromatic transition state in the ethylene-butadiene reaction.[70a]

This is a definite insight, quite remarkable for its time. However, it is questionable to maintain, as some have, that Evans was thus formulating a "rule" of aromatic transition states. Such a rule requires not only that one assert that aromatic transition states are more favorable, but also that *one have a procedure for recognizing which transition states are aromatic.* Without such a procedure, a "rule" of aromatic transition states verges on a tautology. Evans undoubtedly recognized this, and in fact, he attempted to provide a procedure. Unfortunately, his suggestions did not solve the problem. Apparently as a result of an improper conflation of MO and classical VB resonance ideas, Evans had only an uncertain grasp of the concept of aromaticity and especially of antiaromaticity. In this way, he was led to a formulation that stressed the total *number* of reactive electrons:

» *... there are qualitative rules which follow. The energy levels of the mobile electrons lie lower in cyclical structures than in straight chain compounds with the same number of centres available. The energy levels of the mobile electrons are lower the greater the number of available centres. These rules imply that the lowering of the activation energy due to the resonance effect will be greater in cyclisation reactions than in chain formation and that the* *resonance energy in the transition state will increase with the increasing degree of conjugation of the reacting molecules.* «

These rules must be carefully distinguished from the orbital symmetry rules. The overarching lesson taught by orbital symmetry and related theories is that favorable ("allowed") pericyclic transition states *in a given geometry* can be recognized according to whether the number of reactive electrons is ($4N$) or ($4N + 2$). In disrotatory electrocyclic reactions and *supra-supra* cycloadditions, for example, this number is ($4N + 2$), whereas in conrotatory electrocyclic reactions and *supra-antara* cycloadditions it is ($4N$). The predictions thus alternate accordingly. Violation of these requirements results in a "forbidden" reaction.

A literal application of the Evans rule favoring more highly conjugated reactants would predict, for example, that resonance stabilization in the the cyclooctatetraene-like transition state of the hypothetical *supra-supra* [4 + 4] dimerization of butadiene should be greater than that in the benzene-like transition state of the *supra-supra* [4 + 2] association of butadiene and ethylene. But we now recognize the supra-supra [4 + 4] transition state as *antiaromatic* and therefore orbital symmetry forbidden.

Moreover, stereochemistry and electron count are inseparably entwined in orbital symmetry theory. Because Evans had no experimental examples of reactions whose stereochemistry would be favorable with $4N$ electrons, he could not have incorporated them into his thinking except by imagining them, which he did not do. Nothing in his paper gives any indication that he anticipated the idea that is central to orbital symmetry, that of allowed vs. forbidden (aromatic vs. antiaromatic) transition states dependent on orbital phasing. To the contrary, continuation along the direction he started would have ended in a theoretical cul-de-sac. Therefore, Evans's contribution was at most a collateral, rather than a lineal, antecedent in the development of orbital symmetry theory.

On the other hand, the intimate relationship of orbital symmetry and the Hückel "magic numbers" is obvious. Quite analogously to our previous discussion of ground state molecules, it is precisely because of the *cyclic* nature of pericyclic reactions that one expects them to be more properly described by Hückel MO methods than by classical VB theory. The major extension needed for the application of Hückel theory to transition states was the expansion of the original Hückel ($4N + 2$) magic numbers, derived for conventional cyclic conjugated π-electron systems with no forced orbital phase inversions, to another arrangement with a forced phase inversion, for which the magic number is $4N$.

3. 10 Connectivity as a Strong Determinant of Spin in Non-Kekulé Molecules. Violations of Hund's Rule in Biradicals

As we have seen, Lennard-Jones[12] applied to molecules (*e.g.*, O_2, B_2) the generalization, derived from atomic spectroscopy, that later came to be known as Hund's first rule. This governs the prediction of the spin multiplicity in cases where two electrons occupy degenerate orbitals: "that state is held to be lowest which has the greatest multiplicity."[12–15] Some theoreticians choose to interpret Hund's rule in a "strong" form, that is as applicable only when the the orbitals in question are exactly degenerate. Except for rare accidental cases, this would limit the rule to molecules like O_2, in which the degeneracy is symmetry-enforced. Other theoreticians, including Hund himself, use a "weak" form of the rule as applicable even when the degeneracy is only approximate. Indeed, many examples (mostly triplets) exist of high-spin ground states in π-conjugated molecules with only approximately degenerate frontier molecular orbitals (FMOs). This serves to remind us that the physical forces underlying the "strong" form of Hund's rule are present even in the non-degenerate cases, where they may produce a high-spin ground state also.

Obviously, if the energy separation of the FMOs is large enough, it would become advantageous to put both frontier electrons in the lower of the two, producing a singlet ground state. Such cases are known, but it seems fair to say that this kind of molecule does not really violate even the "weak" form of Hund's rule. As we shall see, Erich Hückel in 1936 recognized that a more interesting and subtle situation arises when the connectivity of the π-orbital centers can be considered to result from the union of two radicals at inactive sites, where the NBMO coefficients are zero or nearly zero. The history of this insight and its further development carry a poignant message.

We can take up the story in 1950, when Longuet-Higgins[55c] invoked the approximate form (without so identifying it) in a seminal paper which is perhaps the most influential theoretical writing on π-conjugated biradicals in the period 1930–1970. He showed that, at the Hückel MO theoretical level, non-Kekulé molecules such as trimethylenemethane (TMM), *m*-quinodimethane (MQDM) and tetramethyleneethane (TME)(see Figure 11), each have a degenerate pair of nonbonding (NB) MOs occupied by only two electrons and predicted for each of these molecules a triplet ground state. The symmetry point group of TMM, D_{3h}, contains E-representations, but those of MQDM (C_{2v}) and TME (D_{2h}) do not. Thus, in the latter two cases, the NBMO degeneracies are not symmetry-enforced, and the energies are expected to split apart at higher levels of theory.

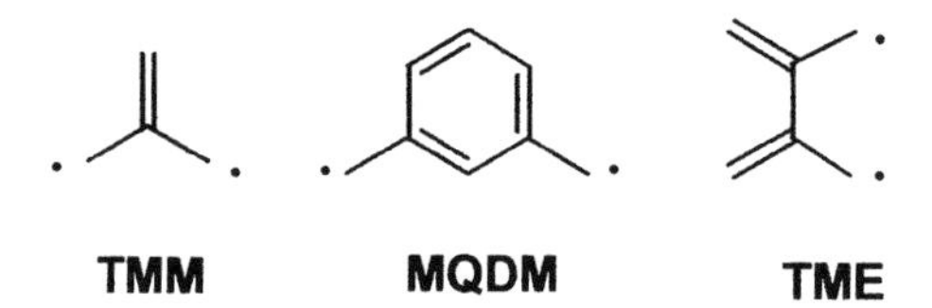

Figure 11. Three non-Kekule molecules discussed by Longuet-Higgins.[55b]

Longuet-Higgins cited no precedent for his predictions of spin multiplicities in organic biradicals, and apparently there had been only one prior such case in the literature: In 1936, Hund himself[71] had predicted a triplet ground state for one of the Schlenk-Brauns hydrocarbons (Fig. 12) studied by Müller et al.[72] This molecule too has degenerate NBMOs at the Hückel level of theory and would have fit easily into the Longuet-Higgins scheme.

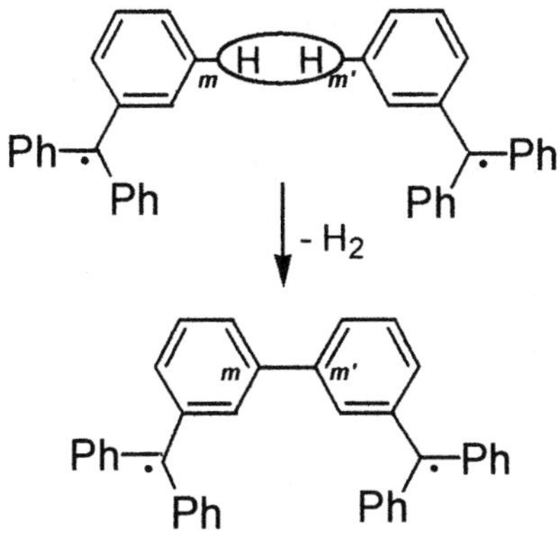

Figure 12. Formal synthesis of the Schlenk-Brauns hydrocarbon as visualized by Hückel.[73]

However, this conclusion was immediately challenged by Hückel,[73a] who pointed out reasons why the rule of highest multiplicity was of dubious validity in the case of the Schlenk-Brauns compound. Hückel's argument developed from his recognition that the Schlenk-Brauns structure could be made (conceptually) by a union of two triphenylmethyl radicals at the *m* and *m'* positions (Figure 12). In the NBMO of triphenylmethyl, the *m* positions of the phenyl rings are nodes, that is, they have π-electron coefficients of zero at this level of approximation. In these circumstances, the exchange energy between the two halves of the Schlenk-Brauns molecule is close to zero. Since it is essentially just the exchange energy that accounts for the separation between the triplet and the lowest singlet, these two states will be almost degenerate. Note that the argument has nothing directly to do with the fact that the NBMO degeneracy in the Schlenk-Brauns hydrocarbon is accidental and will be lifted at higher levels of theory. Instead, in MO terminology, it really is a recognition of the spatial separability of the NBMOs, a property later to be called "disjoint" by Borden and Davidson[74a,75] (see below). Hückel's main conclusion was that the ground state is not predictable at this level of theory, contrary to the previous assumption of Hund's rule.

The Schlenk-Brauns hydrocarbon, one of the few biradicals known in 1936, was an esoteric species at the time. One might then think of Hückel's argument as pertaining to a highly specialized compound of little interest to those outside the small field of biradicals. Indeed the paper excited little or no notice. In my view, however, it had significance far beyond the specific compound, because it showed that the magnetic properties of non-Kekulé *molecules* cannot be treated by imagining that the exchange interaction of two electrons in the field of the nuclei and the remaining electrons will always favor the triplet. On the contrary, such a problem must take into account the *connectivity* of the atoms. This is a profoundly *chemical* point of view, which repeats a rule every chemist knows: molecular structure determines molecular properties.

A simple extension of Hückel's argument would reveal that of the three biradicals TMM, MQDM, and TME (Figure 11), the latter has the same connectivity feature as the Schlenk-Brauns compound (Figure 12): it can be made by a union of two radicals (allyl in this case) at NBMO nodal positions. Applying Hückel's argument then, we can see that TME should not have been classified by Hund's rule.

Because of Longuet-Higgins' (deservedly) high reputation in the field of theory, his unqualified use of Hund's rule in this case had the effect of steering experimentalists along the same oversimplified line of thinking. Also, the absence of a citation to Hückel's 1936 work contributed to decades of obscurity for that paper.[73b] One does not disparage the seminal significance of the work of Borden and Davidson[74a] and Ovchinnikov[74b] by noting that they re-discovered the essence of Hückel's connectivity argument[73a] after a lapse of forty years.[73b] As might be expected after such a passage of time, these papers[74] went far beyond Hückel's original rather qualitative discussion.[73a] They brought to bear the full power of modern electronic structure theory to derive, by both MO[74a] and VB[74b] methods, the idea that when the π-electron system of a biradical has the Hückel type of NBMO node-to-node connectivity, only a small energy gap separates the multiplet states. Ironically, at the time of their publications, these workers also were unaware of Hückel's paper!

Recent experimental and computational confirmations that Hund's rule is violated in properly constituted disjoint π-conjugated biradicals are reviewed elsewhere.[76] The significance of these findings is not so much in the violation of Hund's rule itself. After all, Hund's rule is violated occasionally even in atoms.[14] Rather, it is that a rational basis in theory now exists to guide us to a class of molecular structures in which a violation is likely to occur, and that in the experimental confirmation of this theory, it now has been shown that the *connectivity* of the π-system is a strong determinant of the spin.

3.11 Reflections on Hückel's Career

Hückel's entire contribution to organic quantum chemistry amounted to 17 papers, a list that includes several summaries and reviews. After 1937, his original contributions to this field virtually stopped.[37b] (If Hückel were working now, subject to our dubious emphasis on "productivity," that publication record would put his grant applications in grave jeopardy). The subsequent history of the field of organic quantum chemistry showed that there was much more to be done. What can explain Hückel's withdrawal?

Some have the opinion that he became discouraged when he failed to develop any further ideas in the field comparable in significance to his early work. This may be true, but, of course, it merely pushes the problem one step deeper: Why did he run out of ideas? I cannot claim to have solved this mystery, but some inkling of the truth may emerge from an examination of facts leavened by restrained speculation. It will be convenient to divide the inquiry into scientific and personal factors, although we must keep in mind that, in Hückel's life, as in the lives of most of us, these elements interact strongly, and such a separation is, in the end, artificial and potentially misleading.

3.12 Hückel's Uncomfortable Location and Professional Flaws

By the time he arrived in Marburg in 1937, Hückel already had done his major work in theoretical organic chemistry. He also could have begun to feel, with some justification as we have seen, that the community of chemists was resisting or ignoring his ideas. Not only had he failed to dent the stubborn conservatism of that group, but his location in the physics department at his university did not provide much encouragement for further efforts in chemistry. He soon reverted to physics and tried unsuccessfully to initiate work on the theory of the nucleus, but apparently did not bestir himself to seek professional contact with chemists such as, for example, the talented and creative Hans Meerwein, then professor of organic chemistry at Marburg. (We cannot put all the blame for the missed collaborative opportunities on Hückel, however. For example, in the period around 1950, Meerwein tried to persuade one of his graduate students to undertake the preparation of cycloheptatrienide anion,[77] apparently unaware that a few yards away in the physics department lived a man who twenty years before had predicted the properties of this very species).

In Hückel's failure to make connections with colleagues in chemistry, we see a character trait manifesting itself as a professional shortcoming. To expand his ideas in chemistry, Hückel needed the stimulus of interaction with chemists, but his own shyness held him back. Ideally, his best move might have been to seek a position in a chemistry department, but one doubts that the compartmentalization of disciplines in Germany at that time, which we already have seen, could have made this an easy transition in many universities. Moreover, we must remember Hückel's personal history of long struggle for a permanent job. When a post finally did become available in physics, he did not dare to turn it down.

3.13 Hückel's Personality

But beyond that, there was the problem of Hückel himself. There seems to be agreement among those who knew him that his was a difficult and even paradoxical personality – shy yet sardonically witty, hypochondriacal, petulant, pessimistic, depressed, ultimately lethargic and withdrawn from academic contact with colleagues and students. In the section that follows, we attempt to analyze some of the factors that might have contributed to these characteristics.

A glimpse of Hückel's quirky character comes through in the story of his absence from the Hahn Prize ceremony.[78a] To appreciate the significance of this, we must be aware that formal recognitions of Hückel's work were sparse until late in his career. Some notable distinctions were: in 1965, the Otto Hahn Prize, jointly awarded by the German Chemical Society and the German Physical Society; in 1966, nearly 30 years after his departure, an honorary degree from the Technische Hochschule in Stuttgart; another honorary degree from the University of Uppsala in 1973; election to membership in the Leopoldine Academy in Halle, and in 1977, election as a Foreign Member of the Royal Society.

The Hahn Prize ceremony was scheduled to be held in Bonn in September, 1965, simultaneously with a symposium celebrating the 100th anniversary of Kekulé's benzene formula, but Hückel did not attend. In his letter[78a] to Richard Kuhn, president of the German Chemical Society, Hückel pleads that "the state of my health does not permit me to receive this honor in person," but from his autobiography[78b] we get a significantly different excuse: "I was on vacation and dreaded first the trip and then the pompous style in which the Kekulé symposium was planned. In November the medal was presented to me in a small ceremony, which went very harmoniously and was more to my taste than a big display."

Hückel's personality traits surely contributed to his inability to persuade others of the significance of his scientific work.[77,79a–c] He himself[79a] and several colleagues who knew him have characterized his lectures as sometimes flawed by nervousness and idiosyncrasy. Nevertheless, they were scientifically excellent,[79b,c] and the students loved them.[79b] On occasion, he apparently was capable of delivering a "sovereign and convincing"[79d–f] presentation.

Thus, although Hückel's lectures had substance, I have the impression that he did not have the self-confidence, public presence, or power of articulation that would have been needed to offset the dazzling qualities – expository skill, intellectual breadth, and sheer personal magnetism – of his chief rival, Pauling. The confrontation with Pauling clearly delayed the acceptance of Hückel's ideas.

As we have seen, Pauling's classical VB theory, although successful in other applications, really did not provide a good account of conjugation and aromaticity, yet his ideas dominated the field for decades. Here was a clear case of the power of persuasion. Few organic chemists really knew the theoretical basis of Pauling's ideas on aromatic molecules, but he made it easy for them to apply his theory.

One can only guess what the history of this field might have been had Hückel made a more determined effort to reach his potential constituency in chemistry, for example by casting his equations in the form of the simple mnemonic diagram put forward twenty years later by Frost and Musulin,[80] or even by making use of the catchy ($4N + 2$) slogan. These steps would have gone far to popularize his results and make them more intelligible.

3.14 Illness, Politics, and War. Hückel and the Nazis

The reasons why the characteristics described came to dominate Hückel's psyche remain to be elucidated, but some facts I have found out may be relevant.

Hückel[81] blamed the experiences of the war and a succession of serious illnesses for the exhaustion of his energies. At one time or another, he suffered from a lung abcess, intestinal bleeding, blindness in his left eye, which he ascribed to smoking too many cigarettes, severe migraine headaches, and an episode he called "psychological collapse," which his doctors blamed on high blood albumin.

Beyond these, however, may be another factor that Hückel does not blame directly for his decline but perhaps was as significant as any of these. This was his reluctant involvement with the Nazi Party. In the following pages, I examine the question of whether he carried a burden of self-reproach for this relationship, and whether that may have been a contributing factor in his withdrawal from active research.

At some time between 1933 and 1937 (he does not specify the date), Hückel entered the NSV, the National Socialist People's Welfare organization, an operating arm of the NSDAP, the National Socialist Workers' (Nazi) Party. His reasons for doing this, as described, in his autobiography,[81] are given here in his own words (my translation):

» *With the so-called 'seizure of power'*[82,83] *by Hitler and his hordes, a new problem of existence came upon us. Unlike many 'March violets' [new members who joined the Nazi party when membership was temporarily opened in March, 1933, as contrasted to the 'old fighters,' who had been members before then], I did not enter the Party. I did not know what to do in order to avoid losing my job, which would have become precarious if I did not [join].*

My friend Schumm repeatedly urged me to apply for entrance into the Party. Since I saw no other way to save both my family and my science, I reluctantly decided to apply for membership in the NSV and to take up in it the position of a low-level functionary, a block warden. In this capacity, I had to collect the membership dues, participate in so-called 'indoctrination evenings,' sell 'indoctrination letters,' and on occasional 'collections,' such as those for 'winter relief,' collect on the streets, that is, beg ... I considered the NSV, in these respects, to be harmless, since, as I thought, there at least nothing bad could happen ... After my call to Marburg, I simply let my functionary duties lapse. «

We know now, and Hückel knew in 1975, when he wrote that passage, of the misery, devastation, pillage and murder the Nazis were soon to unleash in Germany and throughout Europe, horrors that still evoke universal revulsion. Hückel's explanation therefore could be interpreted with two extreme hypotheses, with gradations between the extremes. The first is that Hückel was simply being disingenuous here, that he was a dedicated Nazi, politically and personally committed to the cause, who when writing his memoirs in 1975 tried to cover up his past approval of and participation in the Nazis' actions. The second is that, in 1937, he was in fact what he said he was, a marginal figure in German academic life, grateful to have put before him at long last an opportunity to pursue his work and support his family. After having devoted considerable effort to the examination of this question, I am rather strongly inclined toward the second hypothesis, for reasons I now outline.

Three linked questions call for clarification: First, how accurate is Hückel's characterization of the absolute requirement for Party membership to which he says he was subjected in order to obtain his post in Marburg? Second, even if it is true that Hückel was only a reluctant Nazi, was capitulation to coerced involvement his only available course of action? Third, even if the answer to the second question is affirmative, what effect did that capitulation have on his self-regard and ultimately on his work?

Two of the most thorough studies of the relationship between the German academics and the Nazi Party are those of Kelly[84] and Beyerchen,[85] which outline in detail the effects of Nazi policies. These had devastating consequences for the lives and careers of many professors and also for the quality of the universities themselves. They also confronted professors of conscience with painful moral choices.

After their acquisition of power in 1933, the Nazis quickly moved to "purify" the professoriate by promulgating a law that only "Aryans" could hold positions in the civil service. Since all university teachers were civil servants, this meant that the universities would be required to purge "non-Aryan" professors, a category which, within the academic world, connoted largely Jews. This represented a substantial component of the professoriate: In the period 1909–1910, 19% of the teachers at German universities were of Jewish origin, although only 7% held the rank of full professor,[86] proportions that probably did not change greatly before 1933. The impact of the purge was to be especially strong in the natural sciences, where the proportion of Jewish teachers was greater than in other fields. This policy provided the impulse for the departure into exile, over a short period of time, of such "non-Aryans" as Einstein, Franck, Born, and Stern, as well as many others. The debasement of German science caused by these policies was immediate, profound, and persistent.

We should keep in mind, however, that the presence of strong anti-Jewish forces in the German universities was plain well before the Nazis came to power. The experience of the Jewish professor of chemistry, Richard Willstätter, illustrates this. In protest against the actions of his faculty colleagues at Munich in rejecting a series of Jewish candidates for appointment, he had resigned his professorship in 1925. He was soon approached by the University of Heidelberg to consider a professorship there. When word of this leaked out to the newspapers in Heidelberg, students at the university pelted Willstätter with a barrage of hate-filled and threatening letters warning him to stay away. Willstätter regretfully terminated the negotiations.[87]

In parallel with the racist purge, the Nazis soon instituted other policies designed to force the various intellectual disciplines in the universities to emphasize *völkisch* (national-racial) concepts in the education of students. These steps included a variety of mechanisms to ensure the commitment of professors to such concepts, surveillance of the content of courses, and restriction of appointments to politically acceptable candidates. Kelly[84] points out that these policies originated in several different places in the Nazi bureaucracy and never really achieved the universal control that more probably would have resulted had the orders for their implementation come directly from the *Führer* himself. Beyerchen[86b] characterizes the administrative situation generally in the words:

» *The image of the monolith projected by the Nazis has dissolved; the picture is now seen more clearly as a montage of rivalries between mutually antagonistic bureaucratic structures.* «

The enforcement of the Nazi objectives in the universities again was inconsistent. The Party exerted its influence on academic affairs through several means. One was the presence on each campus of *Vertrauensmänner*,[88] trusted Party stalwarts, who served as informers, scrutinizing the activities, writings, and teachings of faculty members and reporting deviant behavior to the authorities. Another was the infiltration or outright takeover of the *Dozentenbund*, the organization of younger teachers in the lower ranks, by Party zealots among the faculty. In some cases, these groups were sufficiently powerful to control recommendations for faculty appointments on campus by threats and intimidation, sometimes (but apparently not always) succeeding in making Party membership a pre-condition for appointment to the faculty.

For example, from data given by Kelly,[89] it can be concluded that during the period 1937–1939, a total of 307 new faculty appointments were made of candidates who were not Party members in 1937. Of these, 250 apparently did not join the Party through 1939. Thus 81% did not seem to find it necessary to join in order to secure their positions.

These and other similar data lead Kelly to conclude that "by 1937, even the *Dozentenbundesführer* (leaders of the teachers' organizations) were making recommendations less on political acceptability and more on academic grounds."[89] Again,[90] there was a "general lack of enthusiasm of the professors for joining the Party and the lack of absolute necessity to do so," and "these figures show that generally an outstanding man could still pursue a successful career in the universities so long as obvious, outspoken political opposition were not charged against him" (and we might add, so long as he was not "non-Aryan").

These conclusions are obviously relevant to the specific issue of whether Hückel's appointment really was, as he claimed, contingent upon Party membership. Thus, in his words,[91] "it was *necessary* for me [emphasis supplied], before I came to Marburg, to join the Party [itself] in 1937. Otherwise, I could not have received my appointment in Marburg, and I would have lost the one in Stuttgart."

Although, as we have seen, no such requirement was *nationally* imposed by the Party leadership, local political ambitions and Nazi zealotry easily could have put it into effect in specific universities. Moreover, the 1933 laws demanded of all civil servants a "guarantee" of loyalty to to the National Socialist state. Membership in the Party itself was widely considered to be a demonstration of such loyalty far stronger than mere membership in the NSV. After all, Hückel could hardly be unaware that at his own Alma Mater, the University of Göttingen, "the political criteria [for appointment to the physics faculty] were basically Nazi party membership and adherence to the attempt by Lenard and Stark to inject racial ideology into physics."[92] My inquiries into local conditions in Marburg[93] suggest that party membership also may well have been an important criterion for appointment there.

It will be useful to give a brief history of the professorship in theoretical physics at Marburg, for which Hückel was a candidate in 1937. The first incumbent of a position in that field was Friedrich Wilhelm Feussner, who originally was appointed as extraordinary professor of mathematics and physics in 1880, became "ordinary honorary professor" in 1908, and took on the title of "leader of the theoretical physics seminars" in 1910, an event which marked the first use of that term on the Marburg faculty. Feussner retired in 1918 and was succeeded by Franz Arthur Schulze in 1919, appointed as extraordinary professor in theoretical physics. Schulze received the title "personal ordinary professor" in 1922 and retired in 1937. It was the position as

Schulze's successor to which Hückel ultimately was appointed, as we have seen, in 1937.

According to the proceedings of the Marburg faculty on the Schulze succession,[93] Hückel was not listed first among the candidates recommended but instead shared second place with two other candidates in a group ranked far below the candidate in first place and listed only in alphabetical order. It was often the case in the faculty deliberations on appointments that political objections were put forward against specific candidates by the *Dozentenbund*, and in this proceeding, such objections were raised against Hückel's competitors. For this reason, it seems very likely that Hückel's existing application for candidacy into the Party may have played an important part in his eventual emergence as the appointee.

Hückel never said who told him that it would be to his benefit to join the Party in order to win the Marburg job, and it may be that he was just responding to rumors and local gossip about how best to proceed. However, his claim of *necessity* is a good deal stronger; it permits the speculation that he may have received information from a knowledgeable informant that such action would give him an advantage over his politically vulnerable competitors.

Note that after 1933, many openings in German academic rosters had been created by the dismissal of "non-Aryan" professors, and candidates to fill them were not in short supply. However, this was not the case for the position Hückel sought. That had been made available by Schulze's normal retirement. Whether Hückel's conscience would have permitted him to take a position opened by the purge is an interesting question to which I have no answer.

3.15 The Ambiguities of Morality

To some, Hückel's acquiescence, however reluctant, to the political realities of that time and place may seem tinged with expediency or even opportunism. We have to realize, however, that he was too insecure to strike a defiant attitude. That would have been a quixotic gesture. His professional credentials did not qualify him then as an outstanding figure in physics, and his job prospects were accordingly limited. Unlike Erwin Schrödinger, a Nobel laureate and one of the very few "Aryan" physicists who resigned his post (at Berlin) in 1933 rather than tolerate the interference of the Nazis, Hückel could not count on being in demand elsewhere. His fate more likely would have resembled that of another "Aryan" young man, Born's chief assistant, Martin Stobbe, who "resigned, destroying his academic career in Germany," because "his conscience could not be reconciled with the demands of the government."[94] Stobbe never found another permanent academic post and apparently died during the war, his promise unfulfilled.

Only a very small number of "Aryan" scientists resigned because of conscience. The case of the couple Berta and Ernst Scharrer[95a] is especially noteworthy. By 1934, Ernst was Director of the Edinger Institute for Brain Research in Frankfurt, and Berta worked there also.[95b]

The Scharrers soon found the political climate in Germay intolerable. By 1937, they had "decided that it was impossible for us to be part of this system any longer."[95a] Through some clever subterfuge, they managed to deceive the authorities and escape to the United States, where the University of Chicago provided them with laboratory space. Thus "in 1937, Berta and Ernst Scharrer left their full life, their friends, and all of their research materials. They came to the United States with nothing but two suitcases, the four dollars each they were permitted out of Germany, and a united clear conscience."[95a] One can only admire the resolve of both of the Scharrers and be inspired by the bittersweet story[95a] of their life in the United States, where, at the Albert Einstein Medical College of Yeshiva University, Berta eventually emerged as a major figure in American biology.

It is my belief that Hückel was a man of conscience, not an ideological Nazi or a racist who could thrive in the poisonous environment of the Third Reich. For example, in his autobiography,[96] he speaks with admiration of his friend Hermann, a crystallographer, later his colleague at Marburg, who had been condemned to death by the Nazis for listening to enemy broadcasts and giving asylum to Jews. Hermann had been in prison awaiting execution but had escaped in the confusion near the end of the war. He came clandestinely to Hückel, who sheltered him until conditions became more stable.

We see this side of Hückel's character again in his relationship to Horst Tietz,[97] now a professor of mathematics at the Technical University of Hannover. Tietz, whose father was Jewish, was of course a "non-Aryan" under the racial law. They both went into hiding until apprehended by the Gestapo in 1943. Tietz's father died in his arms in a Gestapo prison in Kassel, a few hours before Tietz himself was sent to Buchenwald concentration camp. His mother, although an "Aryan," paid with her life, in Ravensbrück concentration camp, for her loyalty to her husband and her hatred of the Nazis. After the war, having survived and been liberated, the orphaned Tietz returned to Marburg, where he became a graduate student in mathematics and eventually Hückel's choice as his first private assistant. Tietz presented a lecture of reminiscence at the memorial meeting for Hückel in Marburg in 1980.[79b] Hückel says[98] of him "Tietz became my most devoted collaborator and best friend, and this friendship carried over to our families." In evaluating what Hückel's real motives were, it seems that, in view of the personal histories, one must give great weight to Tietz's statement: "it is clear to me that Hückel was anything but a Nazi, that he chose a way to ensure his existence and to remain clean."

With this background, we can consider the second of the questions we raised earlier: Could Hückel too, when faced with an agonizing choice, like the Scharrers, somehow have made a life outside of Germany? In a sense, this is an idle question, an experiment we can never do. My intention, however, is not to get the answer but rather to suggest that we try to put ourselves in Hückel's place. His personal circumstances were very different from theirs. He and Anne had three children to raise, whereas the Scharrers were childless. He had finally, at age 41, obtained a real chance at a permanent job. Moreover, one gets the impression from reading Hückel's autobiography that, although he was a man of good will and opposed to tyranny, he did not have the strongly developed political sensitivity, the concern for social justice, or the passionate, change-the-world activism that would have been needed to risk everything in his life for a principle. Or if he did have these qualities, he was either unable or unwilling to articulate them. For example, even in his book, written thirty years after the end of the Nazi era, one never sees him express outrage over the Nazis' systematic destruction of the scholarly tradition of the German universities. Rather, he voices concern for his own family, himself, and his professional field of physics. His contempt and revulsion for the racist "German physics" of Philipp Lenard and Johannes Stark[99] are clear. Yet, writing in 1975, he gives only a muted comment on the fate that befell others, with the words:[100]

» *Each of us had hardships – although we were 'Aryan' – in extracting ourselves from the tyranny of the brown-shirted hordes and in individually holding our own. What the 'non-Aryans' and open opponents of the 'system' must have suffered transcends my powers of imagination.* «

What should Hückel have done when confronted by the choice he had to make? As is often the case, the exigencies of circumstance here cloud the issue of what is "moral" or "principled." More to the point perhaps is the question for each of us: "What would I have done?" I must leave to the answer to the reader's contemplation. In Hückel's mind, the only choice was to do what had to be done to secure his academic appointment.

Now we come to the third question. Suppose, for the sake of argument, that my evaluation of Hückel as an upright, respectable man is correct. Would it not then follow that his decision to acquiesce, however reluctantly or remotely, to an association with forces whose objectives he despised, may have gnawed at his conscience for the rest of his life? Could it have been a contributing cause to his depression and apathy in his later years? Few of us are capable of heroism. Some are tough enough to accept compromises and humiliating capitulations in order to go about living. Others, however, are more sensitive and vulnerable. For a man like that, to commit a transgression of principle, even under the coercion of circumstance, stands out as an unbearably ugly feature of the internal landscape he must view every day. Whether Hückel's sense of his own integrity was damaged by his experience with the Nazis we cannot know for sure, but I think this possibility deserves consideration as a conceivable source of his apathy and declining creativity after his move to Marburg.

3.16 The Response of Academic Institutions to Assaults on Academic Freedom

Hückel's dilemma was made even more acute by the recognition that attempts to resist the encroachment of the Nazis would not draw much support from his faculty colleagues. One might ask whether a unified stance of defiance by the faculty could not have mitigated at least the worst of the damage. Although the question is again unanswerable, it is my opinion, reluctantly reached despite my activist instincts, that to hope for such an outcome probably would have been futile. To explore the reasons for this pessimism, I must make a brief digression.

The professoriate of the German universities in the pre-Hitler period had a tradition of closely guarded faculty autonomy in academic governance.[101] The choice among candidates for new appointments to the faculty, the creation of new institutes to keep up with advances in the various research disciplines, the content of courses, the grading of students, and in fact, all matters pertinent to the academic well-being of the institution had to gain the approval of the faculty. How then could the professors of Germany have permitted the Nazis to make a mockery of their vaunted faculty prerogatives and so quickly destroy the intellectual credibility of their academic enterprise? How could their response have been so abject?

Of course, fear was always present. What professor could ignore the threat of physical violence embodied in the strutting brown-shirts on campus, ready to invade the classroom of any dissident? Who could accept the risk of denunciation by a local *Vertrauensmann*, by the rabidly pro-Nazi German Students Association, or even by a colleague sympathetic to the Nazis? The consequences of individual defiance were known to be often catastrophic. As for collective resistance, that too had little chance of success.

One might dream of an opposition by all the university rectors and all the faculties, in agreement that the universities could not survive as credible institutions if the Nazis had their way, and that the criteria for faculty appointment must be made independent of political and racial considerations. That was only a dream. Apart from the pervasive atmosphere of dread, some professors and rectors saw opportunities for personal advancement. What the Nazis wanted to do provided enterprising careerists with new ways to gain prominence by toadyism. The dismissal of the Jewish professors, under the 1933 law "reforming" the civil service, opened up new faculty appointments to which one could aspire if not troubled too much by scruple. Certainly, there were many German professors whose principles would not allow them to seize these opportunities, but effective opposition would have required that the overwhelming majority maintain that stance.

Moreover, at least in the early stages, much of the Nazis' plan for the universities was met with acceptance if not outright ideological approval by many professors. The German university faculties were traditionally conservative and quietly anti-Semitic.[102] Therefore, although there may have been misgivings over the means, for many professors, the drive to purge the Jews from academic life did not conflict with their values. The German professors' acceptance of the imposition of racial and political criteria for appointment to university faculties was consistent with their traditional feeling of superiority to the rough-and-tumble of democratic politics. In Beyerchen's words,[102]

» *The great majority of scholars viewed the Weimar government with icy reserve: they were willing to serve the German state, but not the Social Democrats. They regarded parliamentary politics as sordid and factional, but they did not realize that their own stance, which was allegedly 'above politics,' was just as divisive as that of the parties they abhorred. The National Socialist German Workers Party of Adolf Hitler was too much a mass movement to attract them. But Nazi rhetoric that rejected the role of a mere political party was appealing. Hitler declared that his movement, too, was above politics. National Socialism promised a national uplifting instead of an international leveling of society.* «

Some of the conclusions of Saul Friedländer's recent study[103] of the period are summarized in a review by Fritz Stern[104] in the words:

» *Mr. Friedländer, in assessing the German response to the anti-Jewish measures, notes – as others have before him – the "moral collapse" of the German elites ... With few exceptions, German academics accepted the dismissal of their Jewish colleagues. Most, indeed, pledged enthusiastic support to the new regime.* «

This climate was not conducive to the formation of a united academic front in resistance to the Nazis. Most relevant to Hückel's situation was the lethargy of the most prominent "Aryan" physicists.[102] Some of the most influential of them found rationalizations for their failure to oppose the regime forthrightly and publicly. In the absence of an organized and principled alternative from the leaders, one cannot be surprised that Erich Hückel, and others like him, capitulated.

One must not assume that the dispiriting lack of courage among the German academic leaders was a historical phenomenon unique to that country. During the 1950's, the universities and other intellectual and cultural institutions in the United States also faced an incursion of the barbarians, egged on by Senator Joseph McCarthy of Wisconsin and other self-appointed monitors.[105–107] Of course, the threat posed by McCarthy was not backed up by a power to do harm comparable to that of Hitler. Nevertheless, McCarthyism was "the most serious attack on academic freedom in American history."[108] By the time the hysteria of the 1950's had subsided, many academic careers and lives had been smashed. Moreover, in the prevailing climate of the "cold war," the stage had been set for a debased electoral politics contested on the issue of the candidates' precise degrees of anti-Communism. And as in Germany, academic freedom in the universities was injured. Again, as in Germany, with few exceptions, the administrators, even of the most prestigious universities,[109] failed to produce leadership of a rational, morally sound opposition. Faculties were too divided in their political views and too feebly committed to real academic freedom to mount an effective protest until long after the the damage was done.

Even the national faculty organization, the American Association of University Professors (AAUP), failed in its duty. This association "was the organization through which the professoriate ordinarily exercised its collective concern for the preservation of academic freedom."[110] Yet, despite strong statements of principle, the AAUP took very little action on specific cases of abuse until it was too late. The reasons for this,[111] although different in detail, shared many elements in common with those that determined the German response. Hence the conclusion[110] that

» *... the Association's failure to perform its expected function had a profound and devastating effect on the academic profession's efforts to combat McCarthyism.* «

Writing of the response of administrators and faculty members to this attack, Schrecker points out[112a] that

» *... these were the men and women who had made a full-time, life-time commitment to the academy. Though they lacked the formal authority of the trustees they nonetheless exercised considerable power and could have, had they wanted to, prevented much of what happened.* «

Yet,

» *few seemed to notice that by putting the jobs of unfriendly witnesses [those who were uncooperative in hearings before the Congressional committees] in question, America's colleges and universities had given Joseph McCarthy and the membership of the House Un-American Activities Committee a say over selecting their faculties.* « [112b]

Universities have sometimes been able to resist attempts to violate academic freedom by officious meddlers among trustees, town residents, and even powerful corporations.[113] However, when the threat comes from intruders like Hitler and McCarthy, who could wield the mailed fist of government power and did not hesitate to do so, the record of the universities is not reassuring. Can academic communities ever reach agreement on the meaning and central importance of academic freedom? Can they find the means to defend it effectively in a crisis? It may be that academic freeedom, by welcoming into the universities a diversity of views and by recognizing the right of professors to pursue their inquiries without outside interference, has created such a mixture of divergent motivations, ambitions, and value systems in the modern faculty that agreement on just what principled opposition *really is* cannot be reached. Is even the thought of this depressing paradox too pessimistic? That question can be answered only by the response of the community to the next tyrant or demagogue who invades the academy.

3.17 Summary and Outlook

In the introduction to his landmark paper on partial valence, Johannes Thiele[114] gives his view of the role of theory in nurturing a fruitful relationship with experiment:

» *The opinions about unsaturated compounds that I shall develop in the following may appear quite rash to many. However, if one holds fast to the idea that a theory actually is nothing other than a point of view which permits known facts to be surveyed in a unified way and new facts to be predicted, a point of view whose value and significance naturally can change with the progress of scientific knowledge, then it seems to me that my opinions satisfy both of these requirements.* «

Certainly, the development of modern electronic structure theory, from the early approaches by Hückel and his contemoparies to the powerful methods of today, illustrates this view. Because the original Hückel MO method was a very approximate form of a theory that others later took to a much higher level, one senses now a certain amused condescension from some adepts, who view it as a quaint relic. One must ask, however, whether any other theoretical advance ultimately has done more to enlighten the thinking of organic chemists than Hückel's brief, bright flare of cognition, regrettably quenched too soon. We find words suitable to his legacy in Eliot's[115] terse justification for homage to pioneers:

» *Someone said: 'The dead writers are remote from us because we* know *so much more than they did.' Precisely, and they are that which we know.* «

3.18 Acknowledgments

Many colleagues have provided relevant advice and documentation. Several have submitted to recorded personal interviews. I am greatly indebted to I. Auerbach, P.D. Bartlett, A.D. Buckingham, K. Hafner, M. Hanack, E. Heilbronner, P.C. Hiberty, E.F. Hilinski, R. Hoffmann, H. Hopf, R. Huisgen, F. Hund, H. Kuhn, W. Lüttke, K. Mislow, T. Nozoe, A. Streitwieser, H. Tietz, H.A. Turner, E. Vogel, W. Walcher, E. Wasserman, and K.B. Wiberg. B.Z. Berson and staff members of the Yale University Library Department of Manuscripts and Archives provided bibliographic help. A shorter version of this article first appeared in *Angew. Chem. Intl. Ed. Engl.* **1996,** *35*, 2750 (*Angew. Chem.* **1996,** *108*, 2922).

3.19 References

(1) Hückel, E. *Ein Gelehrtenleben. Ernst und Satire*, Verlag Chemie, Weinheim, 1975.

(2) Hartmann, H.; Longuet-Higgins, H.C. *Biographical Memoirs of Fellows of the Royal Society*, **1982,** *28*, 153.

(3) (a) Parr, R.G. *Int. J. Quantum. Chem., Symp.* **1977,** 25 gives a theoretician's view of some of the topics covered here. (b) Brief biographical sketches of Erich Hückel are given in: Oesper, R.E. *J. Chem. Ed.* **1950,** *27*, 625, and Frenking, G. *Chem. in unserer Zeit*, **1997**, *31*, 27.

(4) (a) Neidlein, R.; Hanack, M. *Chem. Ber.* **1980,** *113*, I. (b) See also Oesper, R.E. *J. Chem. Ed.* **1950,** *27*, 676.

(5) Debye, P.; Hückel, E. *Physik. Z.* **1923,** *24*, 185, 305.

(6) Longuet-Higgins, H.C.; Fisher, M.E. *Biogr. Memoirs, National Academy of Sciences*, **1991,** *60*, 183. (Memoir of Lars Onsager).

(7) Onsager, L. *Les Prix Nobel en 1968*, p. 169, Stockholm, Norstedt and Söner, 1969; *Science*, **1969,** *166*, 1359.

(8) Hückel, E. (a) *Z. Physik*, **1930,** *60*, 423; (b) *Z. Elektrochem. Angew. physik. Chem.* **1930,** *36*, 641.

(9) van't Hoff, J.H. *The Arrangement of Atoms in Space*, 2nd ed. Longmans, Green, London, 1898.
(10) (a) Pauling, L. *J. Am. Chem. Soc.* **1931,** *53*, 1367, 3225. (b) *Nature of the Chemical Bond*, Cornell University Press, Ithaca, NY, 2nd ed. 1940, Chapter III; 3rd ed. 1961.
(11) Slater, J.C. *Phys. Rev*. **1931,** *37*, 481; *38*,1109.
(12) Lennard-Jones, J.E. *Trans. Faraday Soc*. **1929,** *25*, 668.
(13) Adapted from Coulson, C.A. *Valence*, Oxford University Press, London, 1952, p. 100.
(14) For a review and references, see (a) Kutzelnigg, W. *Angew. Chem. Intl. Ed. Engl*. **1996,** *35,* 573; *Angew. Chem*. **1996,** *108*, 629. (b) Kutzelnigg, W.; Morgan, J.D., III. *Z. Phys. D. Mol. Clusters*, **1996,** *36*, 197.
(15) See also Berson, J.A. in *The Chemistry of Quinonoid Compounds*, (Patai, S.; Rappoport, Z. eds.), Wiley, New York, 1988, Vol. II, Chapter 10, and references cited therein.
(16) Reviewed by: Slater, J. C. *Quantum Theory of Atomic Structure*, McGraw-Hill, New York, NY, 1960, p.282–286.
(17) Mulliken, R.S. *Phys. Rev*. **1933,** *43*, 279.
(18) (a) Penney, W.G. *Proc. Roy. Soc. London*, **1934,** *A 144*, 166. (b) *ibid*. **1934,** *A 146*, 223. (b) For an account of Penney's later career as a leader of Britain's nuclear bomb program, see Lord Sherfield, *Biogr. Memoirs of Fellows of the Royal Society*, **1994,** *39*, 283. Penney served on the target committee for the bombing of Hiroshima and Nagasaki(see ref. 18c) (c) Rhodes, R. *The Making of the Atomic Bomb*, Simon and Schuster, New York, 1986, pp. 522, 626, 628, 677–678.
(19) Kimball, G.E. *J. Chem. Phys*. **1940,** *8*, 188.
(20) Hückel, E. *Z. Physik*, **1932,** *76*, 628.
(21) Hückel, E. *Z. Physik*, **1931,** *70*, 204.
(22) (a) Armit, J.W.; Robinson, R. *J. Chem. Soc*, **1925,** *127*, 1604. (b) For a list of reviews, see, March, J. *Advanced Organic Chemistry*, 4th ed. Wiley Interscience, New York, NY, 1992, p. 40.
(23) (a) Pauling, L. *J. Chem. Phys*. **1933,** *1*, 280. (b) Pauling, L.; Wheland, G.W. *J. Chem. Phys*. **1933,** *1*, 362. (c) Pauling, L.; Sherman, J. *J. Chem. Phys*. **1933,** *1*, 679. (d) Wheland, G.W. *J. Chem. Phys*. **1934,** *2*, 474. (e) Pauling, L. in *Organic Chemistry An Advanced Treatise* (Gilman, H. ed.) 2nd ed. Wiley, New York, 1943, Vol. II, Chapter 26.
(24) (a) Wheland, G.W. *The Theory of Resonance and its Application to Organic Chemistry*, Wiley, New York, 1944.
(25) Wheland, G.W. *Resonance in Organic Chemistry*, Wiley, New York, NY, 1955, p. 632.
(26) (a) Mulliken, R.S. *Ann. Rev. Phys. Chem*. **1978,** *29*, 1, and references cited therein.
(27) Hehre, W.J.; Radom, L.; Schleyer, P. v. R.; Pople, J. A. *Ab Initio Molecular Orbital Theory*, Wiley-Interscience, New York, 1986.
(28) (a) Hiberty, P.C.; Ohanessian, G. S.; Shaik,S.; Flament, J.-P. *Pure Appl. Chem*. **1993,** *65*, 35; (b) Shaik, S.; Hiberty, P.C. *Adv. Quant. Chem*. **1995,** *26*, 99, and references cited therein.
(29) Goddard, W., III; Harding, L.B. *Ann. Rev. Phys. Chem*. **1978,** *29*, 363.
(30) Coulson, C.A. *Valence*, Oxford, 1952, p. 109ff.
(31) Heitler, W.; London, F. *Z. Physik*, **1927,** *44*, 455.
(32) (a) Heisenberg, W. *Z. Physik*, **1928,** *49*, 619; (b) Slater, J.C. *Phys. Rev*. **1929,** *34*, 1293. (c) Cf. Slater, J.C. *Introduction to Chemical Physics*, McGraw-Hill, New York, NY, 1939, p. 367ff.
(33) Bloch, F. *Z. Physik*, **1930,** *61*, 206.

(34) Hückel, E. *Papers and Discussions, Internat. Conf. on Physics*, The Physical Society, London, **1934,** Vol. II, 9.
(35) Slater, J.C. *Papers and Discussions, Internat. Conf. on Physics*, London, **1934,** Vol. II, The Physical Society, London, 1935, p. 53.
(36) Bloch, F. *Z. Physik*, **1928,** *52*, 555.
(37) (a) Hückel, E. *Z. Elektrochem. angew. physik. Chem.* **1937,** *42*, 752, 827. Re-issued as an off-print: *Grundzüge der Theorie ungesättigter und aromatischer Verbindungen*, Verlag Chemie, Weinheim, 1938. (b) *Z. Elektrochem. angew. physik. Chem.* **1957,** *61*, 866. This article, which reviews progress in molecular orbital theory, suggests that Hückel maintained an interest in the subject long after his own creative work on it had ended.
(38) Heilbronner, E. in *Non-benzenoid Aromatic Compounds, Ginsburg, D. ed.*, Interscience, New York, NY, 1959, p. 179.
(39) Streitwieser, A., Jr. *Molecular Orbital Theory for Organic Chemists*, Wiley, New York, 1961, p. 5.
(40) Wiberg, K.B. *Physical Organic Chemistry*, Wiley, New York, 1964, p. 9.
(41) Lowe, J.P. *Quantum Chemistry*, Academic Press, New York, 1978,Chapter 8.
(42) Minkin, V.I.; Glukhovtsev, M. N.; Simkin, B. Ya. *Aromaticity and Antiaromaticity*, (115) Wiley, New York, 1994 p. 116ff.
(43) McConnell, H.M. *Science*, **1996,** *271*, 603.
(44) Kuwajima, S. *J. Am. Chem. Soc.* **1984,** *106*, 6496.
(45) (a) Ref. 30, p. 147ff. (b) Longuet-Higgins, H.C. *Proc. Phys. Soc.* **1948,** *60*, 270.
(46) Maynau, D.; Malrieu, J.-P. *J. Am. Chem. Soc.* **1982,** *104*, 3029.
(47) Mulder, J.C.C.; Oosterhoff, L.J. *Chem. Comm.* **1970,** *305*, 307.
(48) Huisgen, R. tape-recorded interview with Berson, J.A. New Haven, October 29, 1995.
(49) Hückel, W. *Theoretical Principles of Organic Chemistry*, Elsevier, New York, 1958, Vol. I (English translation of *Theoretische Grundlagen der Organischen Chemie*, Verlag Chemie, Weinheim, 7th ed.).
(50) Ingold, C.K. *Structure and Mechanism in Organic Chemistry*, Cornell University Press, Ithaca, NY, 1953, p. 185.
(51) Hammett, L.P. *Physical Organic Chemistry*, McGraw-Hill, New York, 1940, p. 16 ff.
(52) Personal notes by Berson, J.A. on lectures in Advanced Organic Chemistry, a course given by Doering, W. von E. Columbia University, 1946–1947.
(53) Noller, C.R. *J. Chem. Ed.* **1950,** *27*, 504.
(54) (a) Fieser, L.F. in *Organic Chemistry An Advanced Treatise*, H. Gilman, ed. 2nd ed. Wiley, New York, 1943, Vol. I, Chapter 3. (b) See also Fieser, L.F.; Fieser, M. *Advanced Organic Chemistry*, Reinhold, New York, 1961, p. 618.
(55) (a) Coulson, C.A. *Proc. Roy. Soc.* **1939,** *A 169*, 414; (b) Coulson, C.A.; Rushbrooke, G.S. *Proc. Cambridge Phil. Soc.* **1940,** *36*, 193. (c) Longuet-Higgins, H.C. *J. Chem. Phys.* **1950,** *18*, 265. (d) Dewar, M.J.S. *J. Am. Chem. Soc.* **1952,** *74*, 3345, 3357.
(56) (a) Mulliken, R.S. *J. Chim. Phys.* **1949,** *46*, 497. (b) Pullman, B.; Pullman, A. *Les Théories Electroniques de la Chimie Organique*, Masson & Cie, Paris, 1952.
(57) Heilbronner, E. tape-recorded interview with Berson, J.A. Herrliberg, Switzerland, March 30, 1996. Heilbronner reports that in a conversation in 1950–51, Pauling remarked that only two people had ever read the Mulliken paper: Mulliken and the poor guy who had to translate it into French.
(58) (a) Pfau, A.S.; Plattner, P.A. *Helv. Chim. Acta*, **1936,** *19*, 858. (b) Plattner, P.A. *Angew. Chemie,* **1942,** *55*, 131, 154.

(59) Review: Keller-Schierlein, W.; Heilbronner, E. in *Non-benzenoid Aromatic Compounds*, Ginsburg, D. ed., Interscience, New York, NY, 1959, Chapter VI.
(60) (a) Hafner, K. *Angew. Chem.* **1958,** *70*, 419. (b) Ziegler, K.; Hafner, K. *ibid.* **1955,** *67*, 301.
(61) Heilbronner, E. in *Non-benzenoid Aromatic Compounds*, Ginsburg, D. ed., Interscience, New York, NY, 1959, Chapter V.
(62) (a) Nozoe, T. *Seventy Years in Organic Chemistry*, American Chemical Society, Seeman, J.I. ed., Washington, DC, 1991. (b) Nozoe, T. interview with Berson, J.A. Tokyo, May 18, 1995. (c) Nozoe, T. to Berson, J.A. personal letter, June 6, 1995.
(63) Review: Nozoe, T. in *Non-benzenoid Aromatic Compounds*, Ginsburg, D. ed., Interscience, New York, NY, 1959.
(64) Dauben, H.J.; Ringold, H. *J. Am. Chem. Soc.* **1951,** *73*, 876.
(65) Doering, W. von E.; Detert, F.L. *J. Am. Chem. Soc.* **1951,** *73,* 876.
(66) See among others: (a) *Theoretical Organic Chemistry. Proceedings and Discussions of the Kekulé Symposium*, Butterworths Scientific Publications, London, 1959. (b) *Aromaticity, an International Symposium*, The Chemical Society Special Publication No. 21, London, 1967. (c) Non-benzenoid Aromatic Compounds, Ginsburg, D. ed., Interscience, New York, NY, 1959. (d) Garratt, P.J. Aromaticity, Wiley, New York, 1986.
(67) (a) Doering, W. von E.; Knox, L.H. *J. Am. Chem. Soc.* **1954,** *76*, 3203. (b) Breslow, Höver, H.; Chang, H.W. *J. Am. Chem. Soc.* **1962,** *84*, 3168.
(68) (a) Woodward, R.B.; Hoffmann, R. *The Conservation of Orbital Symmetry*, Verlag Chemie, Weinheim, and Academic Press, New York, 1970. (b) Fukui, K. *Acc. Chem. Res.* **1971,** *4*, 57. (c) Zimmerman, H.E. *Acc. Chem. Res.* **1972,** *5*, 393. (d) Dewar, M.J.S. *Angew. Chem. Intl. Ed. Engl.* **1971,** *10*, 761. (e) An important insight undergirding the latter two papers is given by Heilbronner, E. *Tetrahedron Lett.* **1964,** 1923. (f) Longuet-Higgins, H.C.; Abrahamson, E.W. *J. Am. Chem. Soc.* **1965,** *87*, 2045.
(69) Further references: Ref. 68a. p. 169.
(70) (a) Evans, M.G. *Trans. Faraday Soc.* **1939,** *35*, 824. (b) For a much closer early approach to modern orbital symmetry ideas, see Syrkin, Ya. K. *Bull. Acad. Sci. USSR, Div. Chem. Sci.* **1959,** 218. *Izves. Akad. Nauk SSSR, Otdel Khim. Nauk*, **1959,** 238. (c) Syrkin, Ya. K.; Moiseev, I.I. *Uspekhi Khim.* **1958,** *27*, 1.
(71) Hund, F. personal communication to Müller, E., as cited by Müller and Bunge (ref. 72b).
(72) (a) Müller, E.; Müller-Rodloff, I. *Ann.* **1936,** *517*, 134. (b) Müller, E.; Bunge, W. *Ber.*, **1936,** *69*, 2168.
(73) (a) Hückel, E. *Z. physik. Chem. Abt. B*, **1936,** *34*, 339. (b) This paper was first called to my attention in 1978 by E.F. Hilinski, then a graduate student in chemistry at Yale.
(74) (a) Borden, W.T.; Davidson, E.R. *J. Am. Chem. Soc.* **1977,** *99*, 4587. (b) Ovchinnikov, A.A. *Theor. Chem. Acta*, **1977,** *47*, 497.
(75) Similar arguments lead to the idea that square planar cyclobutadiene (not formally a non-Kekulé molecule) also might violate Hund's rule: (a) Borden, W.T. *J. Am. Chem. Soc.* **1975,** *97*, 5968. (b) Kollmar, H.; Staemmler, V. *J. Am. Chem. Soc.* **1977,** *99*, 3586.
(76) (a) Reynolds, J.H.; Berson, J.A.; Kumashiro, K.K.; Duchamp, J.C.; Zilm, K.W.; Scaiano, J.C.; Berinstain, A. B.; Rubello, A.; Vogel, P. *J. Am. Chem. Soc.* **1993,** *115*, 8073; (b) Borden, W.T.; Iwamura, H.; Berson, J.A. *Acc. Chem. Research,* **1994,** *27*, 109, and references cited therein. (c) Wenthold, P.G.; Hrovat, D.A.; Borden, W.T.;

Lineberger, W.C. *Science*, **1996,** *272,* 1456. (d) Matsuda, K.; Iwamura, H. *J. Am. Chem. Soc.* **1997,** *119*, 7412. (e) Clifford, E.P.; Wenthold, P.G.; Lineberger, W.C., Ellison, G.B.; Wang, C.X., Grabowski, J.J.; Vila, F.; Jordan, K.D. *J. Chem. Soc. Perkin Trans 2*, **1998,** 1015.

(77) Hafner, K. tape-recorded interview with Berson, J.A. Darmstadt, April 3, 1996.

(78) (a) Undated letter of Hückel, E. to Kuhn, R. reproduced in the *Program of the Kekulé Celebration*, Bonn, 1965. (b) Ref. 1, p. 138 ff.

(79) (a) Ref. 1, p. 134. (b) Tietz, H. "Einige persönliche Worte zu Erich Hückel," unpublished text of a celebratory colloquium, presented in Marburg, November 14, 1980. (c) Walcher, W. letter to Berson, J.A. March 6, 1996. (d) This characterization of Hückel's review of molecular orbital theory in a lecture before the Bunsen Society (ref. 79e) is given by a listener on that occasion (ref.79f). (e) Hückel, E. *Zeit. für Elektrochem.* **1957,** *61*, 866. (f) Lüttke, W. personal communication to the author, July 7, 1998.

(80) Frost, A.A.; Musulin, B. *J. Chem. Phys*, **1953,** *21*, 572.

(81) Ref. 1, p. 138ff.

(82) Hückel's way of referring to the political upheaval that made Hitler Chancellor of Germany in 1933 suggests a certain skepticism over whether that is the right nomenclature. We have no further evidence of his attitude on this question, but it is significant that some recent historical scholarship reaches a similar conclusion. Thus, Turner's study (ref. 83) makes a case that Hitler's accession was by no means inevitable and that …
»*as with so much in the mythology of his Third Reich, the belief that January 30*th*, 1933, marked a seizure of power was spurious. In reality, Hitler had not seized power; it had been handed to him by the men who at that moment controlled Germany's destiny.*«

(83) Turner, H.A. *Hitler's Thirty Days to Power: January 1933*, Addison Wesley, New York, NY, 1996, p. 161.

(84) Kelly, R.C. *National Socialism and German University Teachers: The NSDAP's Efforts to Create a National Socialist Professoriate and Scholarship*. Ph. D. Dissertation, University of Washington, 1973, University Microfilms 73-13 845, Ann Arbor, MI.

(85) Beyerchen, A.D. *Scientists Under Hitler: Politics and the Physics Community in the Third Reich*, Yale University Press, 1976.

(86) (a) Ref. 85, p. 5. (b) Ref. 85, pp. ix–xx.

(87) Willstätter, R. *Aus Meinem Leben*, Verlag Chemie, Weinheim, 1949, p. 346ff.

(88) Ref. 84, p. 211.

(89) Ref. 84, p. 305. This source states that in the period 1937–1939, a total of 463 faculty appointments were made in Germany. Of these, 213 were party members, and of that number of 213, 156 had joined the party before 1937, leaving 57 who joined after 1937, when membership was opened again. Thus, only 57 of the (463 – 156 =) 307 new appointees who were not Party members in 1937, or 19%, felt it necessary to join the Party in order to obtain their appointments.

(90) Ref. 84, pp. 375–376.

(91) Ref.1, p. 140.

(92) Ref.85, p. 9.

(93) (a) I am much indebted for helpful correspondence on these points during April–May, 1996, with Professor W. Walcher, emeritus professor of physics at the university, and with Professor I. Auerbach, head archival consultant at the Hessiches Staatsarchiv Marburg. (b) As kindly abstracted by Professor Auerbach.

(94) Ref. 85, p. 30.

(95) (a) Purpura, D.P. *Biogr. Mem.Nat.Acad.Sci*. **1998,** *74*, 289 (memoir of Berta V. Scharrer). (b) Although a highly trained and competent investigator in her own right, Berta could not be paid under the strict nepotism rules of the time.
(96) Ref. 1, p.158.
(97) (a) Tietz, H., personal correspondence with Berson, J.A., July, 1998. (b) I am greatly indebted to Professor Tietz for the communication (ref. 97a) of relevant aspects of his personal history and for permission to retell them here.
(98) Ref. 1, p. 146.
(99) Ref. 1, pp. 78–79.
(100) Ref. 1, p. 141.
(101) Ref. 85, p. 1ff.
(102) Ref. 85, pp. 4–5.
(103) Friedländer, S. *Nazi Germany and the Jews*. Vol. I: *The Years of Persecution*, 1933–1939. Harper Collins, New York, NY, 1997.
(104) Stern, F. *The Worst Was Yet to Come*, in *New York Times Book Review*, p. 12, February 23, 1997.
(105) For histories of those events, see refs. 106 and 107.
(106) Schrecker, E. W. *No Ivory Tower: McCarthyism and the Universities*, Oxford University Press, New York, NY, 1986.
(107) Caute, D. *The Great Fear: The Anti-Communist Purge Under Truman and Eisenhower*, Simon and Schuster, New York, NY, 1978.
(108) (a) Ref. 106, p. 315. (b) Ref. 106, pp. 315–328.
(109) Hershberg, J.G. *James B. Conant: Harvard to Hiroshima and the Making of the Nuclear Age*, Alfred A. Knopf, New York, NY, 1993, p. 606ff.
(110) Ref. 106, p. 315.
(111) Ref. 106, p. 315–337.
(112) (a) Ref. 106, p. 11. (b) Ref. 106, p. 215.
(113) (a) Schrecker, E. *Academe, Bulletin of the American Association of University Professors*, **1998,** *84*, 8. (b) Pollitt, D.H.; Kurland, J.E. *Academe, Bulletin of the American Association of University Professors*, **1998,** *84*, 45.
(114) Thiele, J. *Ann*., **1899,** *306*, 87. I thank Prof. R. Huisgen for this citation, which appears epigraphically in his own doctoral dissertation.
(115) Eliot, T.S. Tradition and the Individual Talent, in *The Sacred Wood*, Methuen, London, 1928, p. 52.

Chapter 4

The Dienone-Phenol Mysteries[1a]

"There are no general reactions"[1b]

4.1 Introduction

"Explosion" has become a commonplace metaphor for the rapid expansion of organic chemistry and the increasing separation of its several sub-disciplines. Propelled irresistibly along the narrow trajectory of our own specialty, we may lose sight of the historical fact that the branches of our subject have common origins. Yet, the mutual enrichment of the fields of synthesis, natural products, and reaction mechanisms has been characteristic of organic chemistry from the beginning and remains so today. This chapter describes one example, from many in the literature. The story begins with the isolation and structural definition of certain important physiologically active pure substances from natural sources. These achievements generated a strong motivation toward synthesis, which in turn, led to deep questions about the mechanisms of certain molecular rearrangements and the problem of predictability in chemistry.

4.2 Isolation of the Estrogens

In 1927, Adolf Butenandt, a student of Adolf Windaus[2a] at Göttingen, began work on the isolation of the estrogenic hormones from natural sources. He recalls the events in a letter, reproduced verbatim in the original German in the Fiesers' book on steroids,[2b] excerpts from which I give here in English.

» *On the advice of Windaus, I began working on the sex hormones in the year 1927... It is true that in 1927, the Schering Company invited Windaus to work on the follicular hormone. However, because he wanted to dedicate himself to his researches on Vitamin D, he left this collaboration with Schering to me, for which I have always been especially grateful. Then I worked for a year attempting to prepare the follicular hormone from extracts of placentas and first began to work on urinary fractions in 1928 ... In the first years, I carried out the work on the sex hormones alone, helped in the animal researches by Erika von Ziegner, who came to me as a technical assistant at the beginning of the hormone work and whom I married in the year 1931, just after my habilitation. After the preparation of estrone (Scheme 1)*[3] *in pure form, which was our good fortune to achieve together in 1929, several young students came as co-workers in the hormone field.* «

Scheme 1

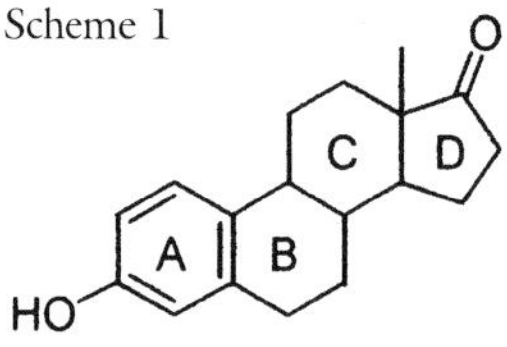

1, estrone 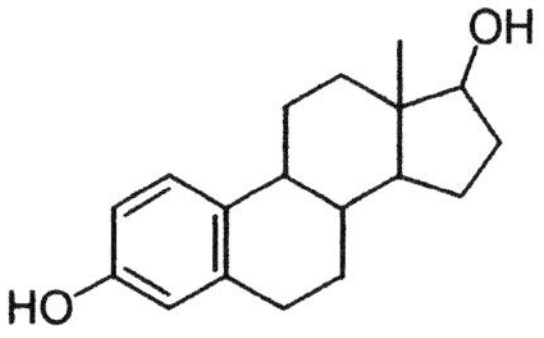 **2, estradiol** 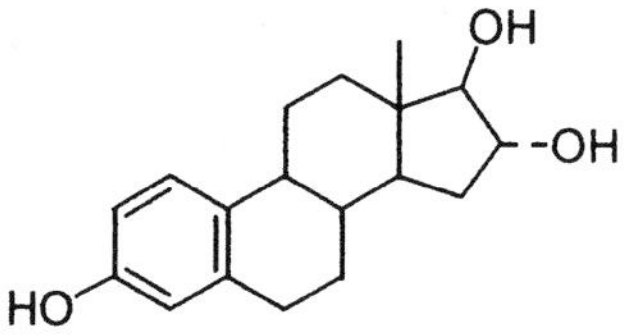 **3, estriol**

Following this discovery of estrone **1** and those of the other female sex hormones, estradiol **2** and estriol **3** (Scheme 1),[3] the remediation of certain human female reproductive malfunctions through hormonal therapy became an important goal. However, the natural hormones were available in only limited amounts, and a search began for sources other than mammalian pregnancy urine.

4.3 Approaches to Synthetic Estrogens by Aromatization of Ring A Alicyclic Steroids

Although the estrogens, like other such naturally occurring steroids as ergosterol, stigmasterol, cholesterol, and the androgenic hormones (Scheme 2), embody the characteristic tetracyclic nucleus, there is a major difference in that ring A of the estrogens is aromatic. This suggested that it might be possible to prepare estrone and other ring A aromatic relatives by aromatization of naturally abundant steroidal precursors, the two most readily available being cholesterol, which is widely distributed in the animal kingdom, and stigmasterol, a component of common plant oils, especially that from soy beans (Scheme 2).

Scheme 2

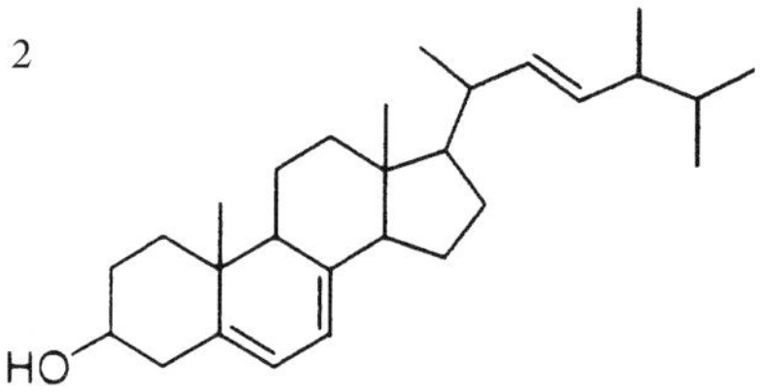

4, ergosterol 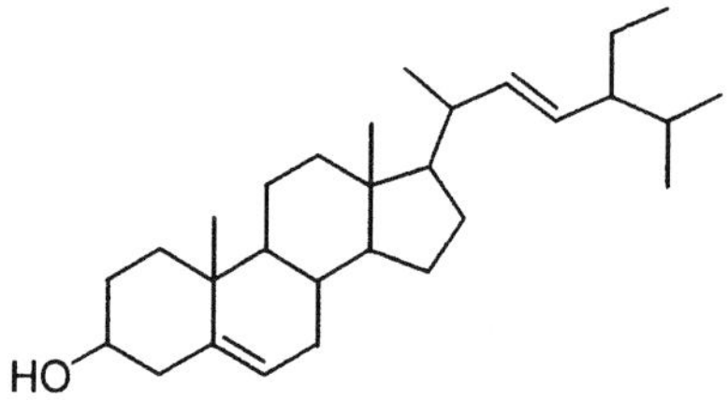 **5, stigmasterol**

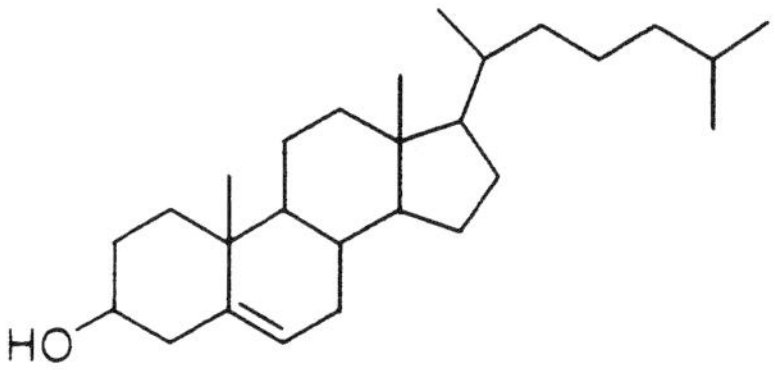

6, cholesterol 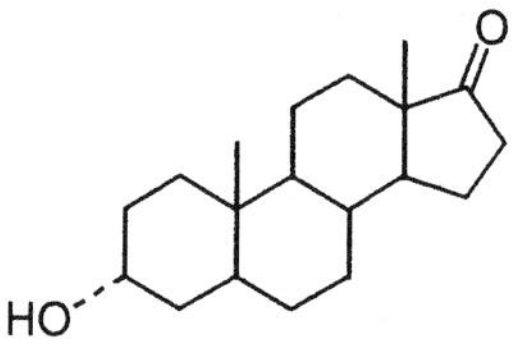 **7, androsterone**

4.3.1 By Pyrolysis

One approach to these goals originated in the Ph. D. thesis work of Hans Herloff Inhoffen, a later Windaus student, who remembers[4a] that:

» *The first and fundamental stimulus to working out the theme of aromatization was given by the closer examination of the transformation of ergopinacone to neoergosterol, one of the first problems that was proposed to me by Prof. Windaus in 1932. At that time the follicular hormone was not known to be a phenolic steroid; only later did the goal of preparing it by partial aromatization of the steroid tetracyclic system give the problem its special direction and significance. When I left Göttingen, I was allowed to take this subject with me for further development; also for this I here must express again my heartiest thanks to my highly honored teacher.* «[4b]

Briefly stated, the work of Inhoffen and others[5] showed that "ergopinacone" **8**, which embodies 1,4-cyclohexadiene units in both ring B sites, undergoes the remarkable reaction shown here (Scheme 3).

Scheme 3

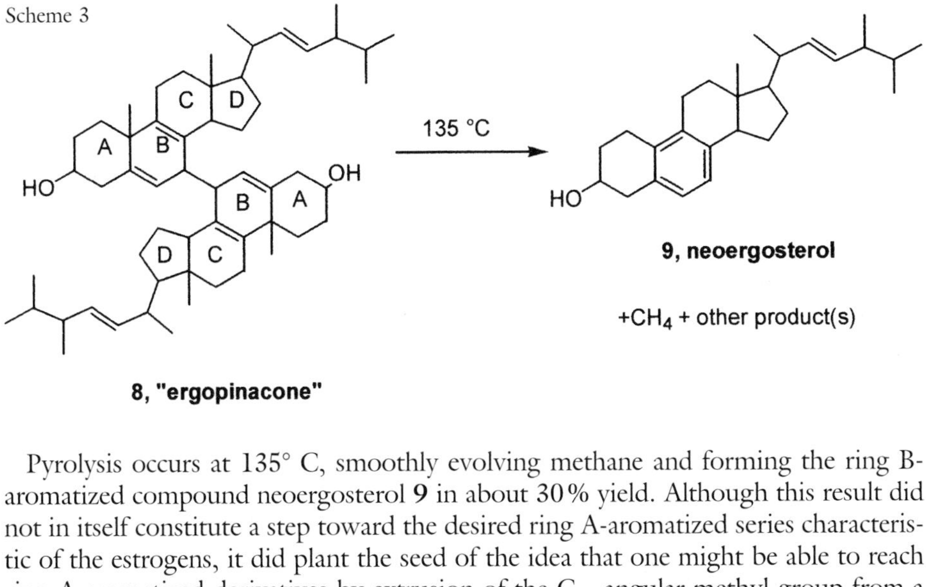

8, "ergopinacone"

9, neoergosterol

$+CH_4$ + other product(s)

Pyrolysis occurs at 135° C, smoothly evolving methane and forming the ring B-aromatized compound neoergosterol **9** in about 30% yield. Although this result did not in itself constitute a step toward the desired ring A-aromatized series characteristic of the estrogens, it did plant the seed of the idea that one might be able to reach ring A-aromatized derivatives by extrusion of the C_{10} angular methyl group from a ring A 1,4-cyclohexadiene. Inhoffen pursued this project in Berlin at the main laboratory of Schering A.G. In fact, although 1,4-cholestadien-3-one (**10**, Scheme 4) gave a mixture from which the desired phenol **11** could not be separated in pure form, the goal was reached[5] when the pyrolysis (at 325° C) of 1,4-androstadien-3-one-17-ol (**12**, Scheme 4) gave a 5% yield of pure estradiol (**2**). Since the androstadienolone **12** can be prepared from cholesterol by a lengthy sequence, the overall process amounts to a formal partial synthesis of estradiol **2** from a commonly available steroid.[6]

Scheme 4

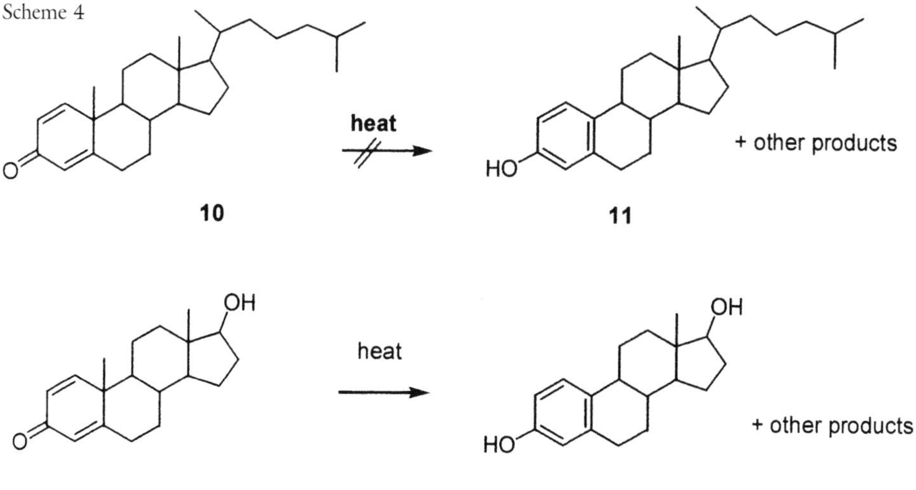

10 **11**

12 **2, estradiol**

4.3.2 By Dienone-Phenol Rearrangement

By 1938, Inhoffen and Huang-Minlon[7] had conceived the idea of a ring A aromatization in a steroidal 1,4-dien-3-one by an acid-catalyzed rearrangement, in analogy to the known[8] aromatization in the rearrangement, apparently by methyl migration, of the sesquiterpene santonin **13** to desmotroposantonin **14** (Scheme 5).

Scheme 5

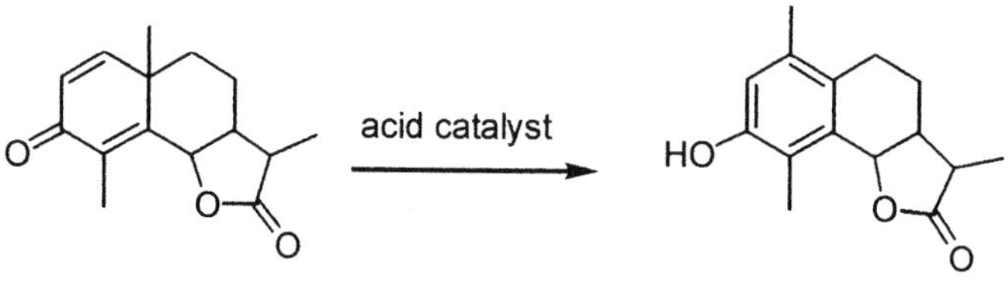

13, santonin **14, desmotroposantonin**

An analogous rearrangement in 1,4-cholestadienone-3 (**10**) would give the 1-methyl-3-tetralol (**15**, Scheme 6) containing the phenolic unit characteristic of the estrogens. Inhoffen and Huang-Minlon[7] indeed found that treatment of the dienone **10** with acetic anhydride/sulfuric acid followed by saponification of the derived acetate gave a 90% yield of a crystalline substance isomeric with **10** to which they assigned the structure **15**. Inhoffen and Zühlsdorf[9] carried out a similar reaction on the androstadienolone **12** and obtained what they believed to be a 1-methylestradiol **16** (Scheme 6).

Scheme 6

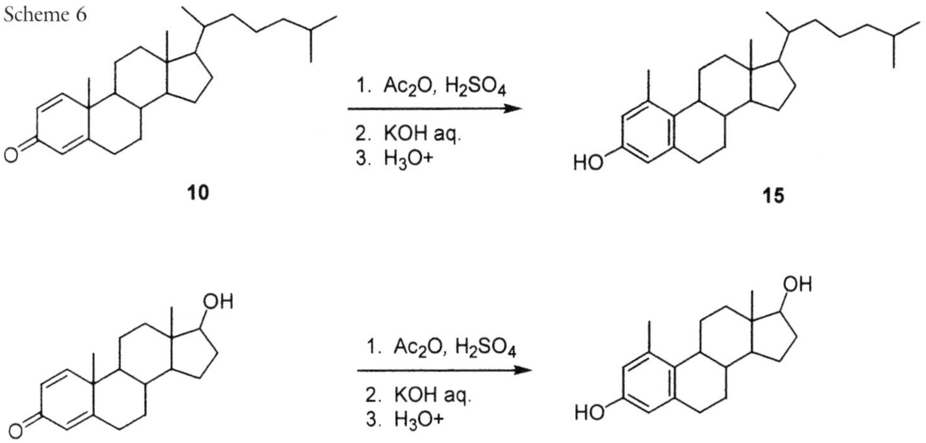

1. Ac_2O, H_2SO_4
2. KOH aq.
3. H_3O+

HO

10 **15**

OH OH

1. Ac_2O, H_2SO_4
2. KOH aq.
3. H_3O+

HO

12 **16**

Structures of **15** and **16** are those

originally assigned[7,9]

Surprisingly, the phenolic product, supposedly **15**, from cholestadienone **10** was cryptophenolic (insoluble in dilute aqueous alkali), and the phenol, supposedly **16**, from androstadienolone **12** was inactive in the Doisy-Allen biological assay for estrogenic activity. The latter result was particularly disappointing, since other 3-hydroxy ring A aromatic steroids had been shown to be active.

Further examples of such rearrangements of steroids were reported in the immediate post-World War II period.[10,11] As in the cases of the products from **10** and **12**, the structures were assigned largely by analogy to the case of santonin. Independent structural elucidation in that era, before the advent of nuclear magnetic resonance (NMR) spectroscopy, would have been a daunting task, and no such confirmation was provided An especially important model (non-steroidal) case was the rearrangement of the synthetic tetracyclic dienone **17** to the phenol **18** (Scheme 7).[12] Unlike those of the steroid cases, the structure of the product **18** was unambiguously established by independent synthesis. In this model system, therefore, methyl migration did occur, which seemed to provide strong support for the methyl migrations proposed in the steroid cases: **10** → **15, 12** → **16** (Scheme 6)[9] and others.[10]

Scheme 7

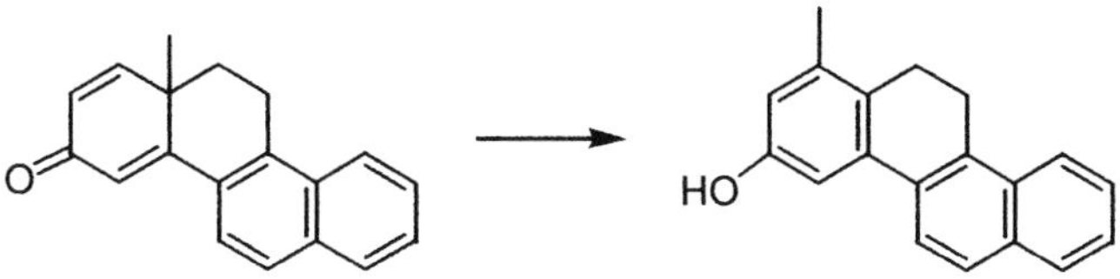

HO

17 **18**

4.3.3 Woodward's Challenge

Nevertheless, soon afterward, the whole basis of these assignments was called into question by Robert Burns Woodward (Figure 1) and his student Tara Singh.[11] In 1950, Woodward was 33 years old, although already recognized as a rising star. From a viewpoint a half-century later, with the knowledge of Woodward's subsequent brilliant achievements, the present-day observer may feel no surprise that this particular newcomer would boldly challenge the conclusions of seasoned, highly competent investigators of these rearrangements. Nevertheless, to those of us who followed the field at the time, his audacity was breathtaking. As events would soon prove, his ideas, although corrrect in a limited sense, did not nearly encompass the extraordinary structural complexity and mechanistic diversity of the dienone-phenol rearrangement.

Figure 1. Robert Burns Woodward (1917–1979), Donner Professor of Science, Harvard University. The image is reproduced from a brochure distributed at his memorial service, held on November 9, 1979 at the Harvard Memorial Church, Harvard Yard, Cambridge, Massachusetts. From the private collection of the author. Reproduced with the kind permission of Ms. Crystal Woodward. Photo by Fabian Bachrach.

What led Woodward to undertake an investigation of the dienone-phenol rearrangement? This question comes in two parts: First, why did he not accept the structures assigned to the steroidal rearrangement products? Second, why did he consider it necessary or even worthwhile to study the issue?

4.3.4 Misgivings about the Structures

Woodward and Singh[11] were the first to note that the structures proposed, **15** and **16** (Scheme 6), were inappropriate for the properties[7,9] of the steroid-derived phenolic rearrangement products. To them, the absence of estrogenic activity, described above, apparently was the most compelling discrepancy, but also, the cryptophenolic nature was unexpected for the supposed structures, which are phenols with two free ortho positions. To provide direct evidence of an *o,o'*-di-unsubstituted phenolic unit in the rearranged phenols, Inhoffen and Zühlsdorf[9] had attempted to prepare a dibromide from the rearrangement product they called **16**. They obtained only a *mono*bromide, rather than the dibromide expected. Attempting to force reaction at the supposedly remaining *ortho* position, they carried out a diazo coupling reaction on the monobromide. Coupling occurred, but only with *concomitant debromination*. As Woodward realized, this was behavior expected of a phenol with only one free *ortho or para* position.[11] Thus, the data seemed to call for alternative structures for the rearrangement products. Apparently Woodward's conviction was strong enough for him to proceed according to this hypothesis, even in the face of the well-established[12] example **17** → **18** (Scheme 7). In the case of the steroidal phenols, the facts themselves spoke louder to him than analogy with a model compound.

4.3.5 Why Did Woodward Undertake the Correction of the Phenolic Structures?

The literature is full of papers in which one suspects that a structure may be wrong. What impels an investigator to choose one of them to re-examine?

The determination of chemical structures was a major undertaking for organic chemists of that era. Often it involved tedious degradation of the molecule to identifiable derivatives, or in some cases, independent synthesis. Unless one had some quick, clever way to prove the published structure wrong, the would-be corrector of a structure might well hesitate to undertake the task. Of course, occasionally an alternative to a suspectedly incorrect structure would suggest itself and could be tested rapidly. Some young organic chemists of that era, especially those under pressure to establish academic credentials for promotion to tenure, bolstered their publication records by scouting the literature for such cases. These "quickie" exercises were not necessarily trivial. Often, the incorrect structure in the literature dated from an earlier time when mechanistic understanding was primitive. The correct answer often emerged from an analysis of the reaction from the modern mechanistic point of view or from the application of the new instrumental techniques such as ultraviolet-visible (UV-Vis) or infrared (IR) spectroscopy. In these skills, the younger, more recently trained chemist was likely to be far superior to the classically trained predecessor. Each such correction served to focus attention on the power of mechanistic reasoning and modern instrumentation in chemistry. In some cases, new aspects of mechanism were brought to light.

Such studies often were not part of a comprehensive research scheme, and they led to no further development. Nevertheless, even if not especially admirable, the practice was understandable and not unusual. In fact, the present writer cannot entirely escape censure for occasional youthful indulgence in such activity. Is it possible that the motivation for Woodward's involvement in the dienone-phenol rearrangement had no deeper origin than this?

In my judgment, the answer is clearly "no." First, Woodward's history must be kept in mind. Despite his youth, by 1950, when he took up the dienone-phenol problem, he had been an independent researcher at Harvard for 13 years. He already had major achievements to his credit, including the valuable additivity rules for UV-Vis spectroscopy, the deduction of the structures of penicillin and strychnine, a critical contribution to the mechanism of the Diels-Alder reaction, and the ground-breaking total synthesis of the important alkaloid quinine. He was supervising a large and active research group. Clearly, mere corrections of structural misassignments in the literature could not be allowed to distract his attention from matters of greater consequence.

As we shall see, there was a strong mechanistic motivation to study the dienone-phenol rearrangement (see Section 4.5.1), but before that, we make a brief digression to demonstrate another motivation, namely that the dienone-phenol matter was directly related to one of Woodward's biggest research projects, the total synthesis of steroids.

4.4 Woodward and the Total Synthesis of Steroids. Targets, Approaches, and Achievements.

In the period 1932–33, the Girard group in Paris13 reported the isolation of two further estrogenic hormones, equilenin **19** and equilin **20** (Scheme 8), from pregnant mare's urine. The structures were established in the mid-1930s by several research groups.[13] Although these hormones are weaker estrogens than estrone **1** and estradiol **2**, they soon acquired an independent significance: One now might hope to achieve the total synthesis of a natural steroid even with the relatively unsophisticated methods of the time. Especially equilenin **19** (Scheme 8), with rings A and B both aromatic, and hence with only two stereocenters to be controlled, posed a greatly simplified objective compared even to estrone, with four centers, let alone to androsterone, cholesterol, and other hydroaromatic steroids, which have six or more such centers. In fact, it now became possible to imagine progress through a series of steroids increasingly hydrogenated and hence increasingly difficult to synthesize: stage I, the naphthalenoid equilenin **19**; stage II, the benzenoid estrone **1**; and stage III, the hydroaromatic androsterone **7** (Scheme 8). This intellectual challenge, as well as the potential medicinal significance of these materials and their scarcity in nature, gave further impetus to synthetic approaches. By the end of the 1930s, the achievement of this sequence of objectives had become the motivating principle of an enterprise which engaged the passionate commitment of some of the world's best synthetic chemists for the next two decades and beyond.

Scheme 8

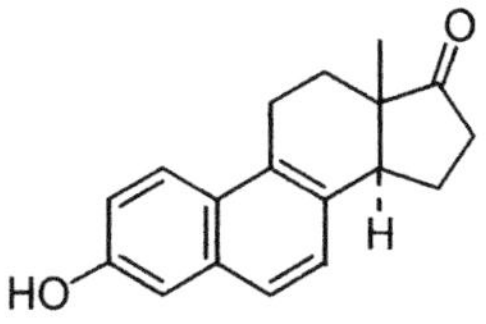

19, equilenin

STAGE I

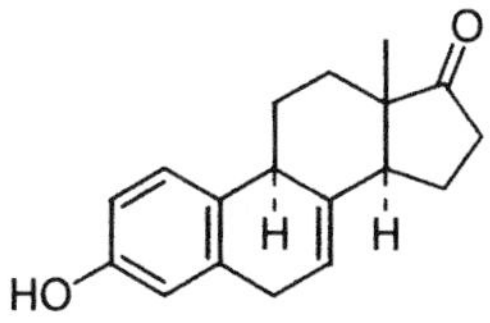

20, equilin

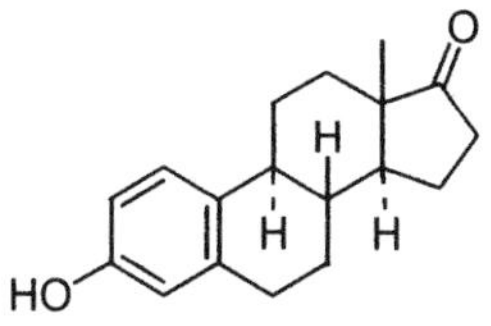

1, estrone

STAGE II

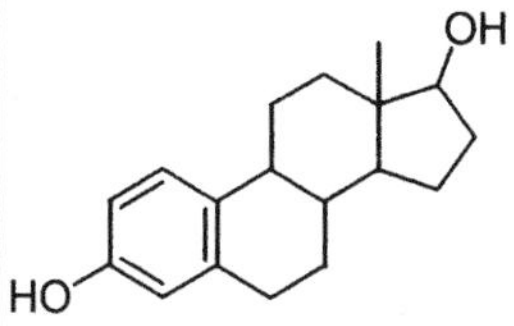

2, estradiol

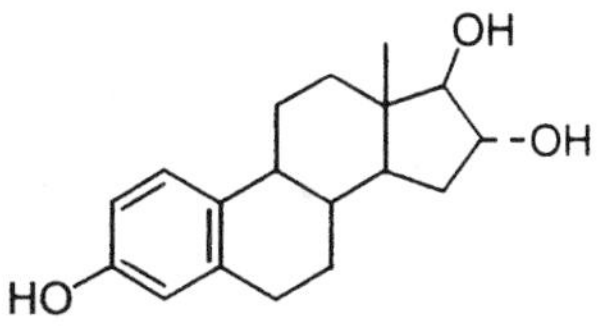

3, estriol

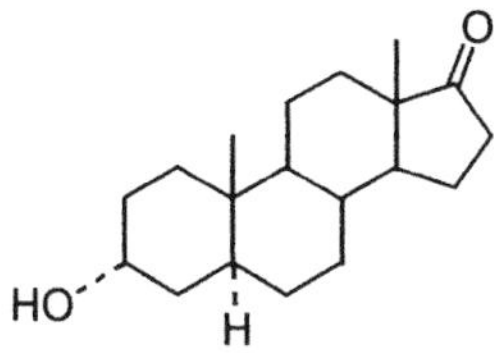

7, androsterone

STAGE III

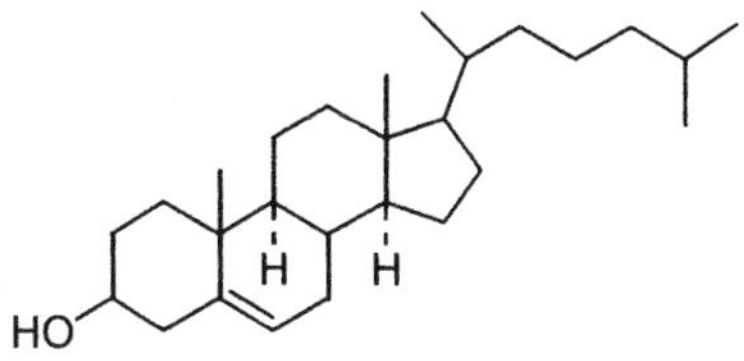

6, cholesterol

4.4.1 Stage I. Rings A-B Aromatic Steroids

The first goal was reached in 1939 by Bachmann, Cole, and Wilds,[14] who effected a total synthesis of racemic equilenin **19** (Scheme 8) in fourteen steps from a synthetic tricyclic naphthalenoid ketone. After resolution as the menthoxyacetic ester, the synthetic equilenin proved to be identical with the natural product from mare's urine. For the most part, the individual steps were based on well known synthetic reactions, and the design was essentially linear rather than convergent.[15] Hence the overall synthesis might be considered uninteresting by today's standards, which emphasize novelty and efficiency. However, by the standards of 1939, the work must be considered brilliant and path-breaking. The authors recognized and overcame, really for the first time, several crucial strategic requirements, not the least of which was the maintenance of the supply of intermediates through the late stages. These challenges still are common to all multi-step total syntheses.

4.4.2 Stage II. Hydroaromatic Steroids

Direct documentation exists that Woodward's interest in estrone and the hydroaromatic steroids already was strong before 1940. In fact, it is hard to avoid the conjecture that, inspired by Bachmann's success with equilenin in 1939, and driven by the challenge of the much more difficult estrone, the 22-year-old Woodward must have been eyeing this target soon after. Two Woodward papers of 1940, one of which was based on his Ph. D. thesis, show that he had already embarked on the path to estrone,[16] and the general motivation is given in Woodward's 1937 Ph. D. thesis:[17]

» *The determination by analytical methods of the probable structures of the sterols, hormones and other naturally occurring phenanthrene derivatives has been the triumph of organic chemistry in the past decade. The problem must be considered unfinished, however, until the present views have been confirmed by the total synthesis of one or more of these important substances. It is not surprising that a number of laboratories have initiated the attack on the problem.* «

Since Woodward carried out his doctoral studies without a conventional mentor,[18] this statement can be taken as defining his own view.

In the Fall of 1941, Harry Wasserman, then an undergraduate at the Massachusetts Institute of Technology and a candidate for graduate study at Harvard, visited Woodward there to discuss research projects, among them, estrone.[19] Wasserman was to become one of Woodward's first graduate students. The written record[20] shows Woodward outlining the hierarchy of difficulty in the sequence equilenin-estrone-androsterone (stages I-III shown in Scheme 8) and explicitly referring to Bachmann's work. Moreover, it seems likely that even at this early stage, his ambition extended beyond estrone to the hydroaromatic steroids themselves. His notes[20] contain the formulas redrawn here (with formula numbers added) in Scheme 9.

Scheme 9

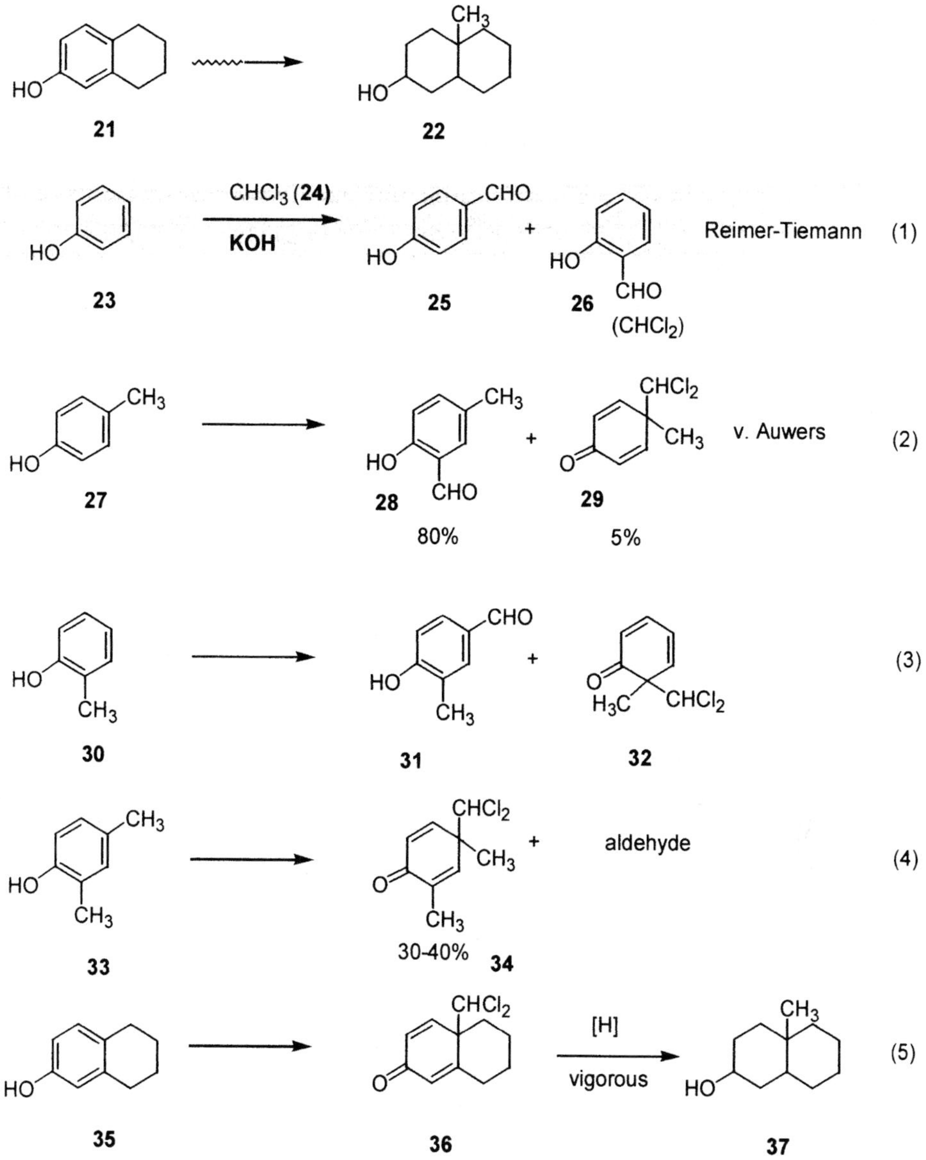

CH_3
HO **21** → HO **22**

$CHCl_3$ (**24**), KOH
HO **23** → HO **25** CHO + HO **26** CHO ($CHCl_2$) Reimer-Tiemann (1)

HO **27** CH_3 → HO **28** CHO CH_3 80% + O **29** $CHCl_2$ CH_3 5% v. Auwers (2)

HO **30** CH_3 → HO **31** CHO CH_3 + O **32** H_3C $CHCl_2$ (3)

HO **33** CH_3 CH_3 → O **34** $CHCl_2$ CH_3 CH_3 30-40% + aldehyde (4)

HO **35** → O **36** $CHCl_2$ → [H] vigorous → HO **37** CH_3 (5)

This is obviously a carefully reasoned attempt to persuade Wasserman that the problem of constructing an angularly methylated fused bicyclic system, so crucial in the synthesis of hydroaromatic steroids, might be solved by application of the studies of von Auwers on the "abnormal" Reimer-Tiemann reaction. One can only marvel at the cool self-confidence of the young Woodward, a freshly appointed instructor at Harvard, who had not yet collected a cadre of co-workers around himself, nevertheless proposing to engage in direct competition with Bachmann, the hero of the first steroid synthesis. As it happened, Bachmann already was well advanced toward the eventual syntheses of "estrone-A",[21] a stereoisomer of estrone. This synthesis and those of estrone itself by Anner and Miescher[22] at CIBA AG in Basel and by Johnson and co-workers at the University of Wisconsin in Madison[23] solved the angular methylation problem by more straightforward but nevertheless effective methods. Even in Woodward's eventual synthesis of hydroaromatic steroids,[24] the angular methyl groups were introduced by original but different devices.

One may conjecture that Woodward soon realized that the dienone approach, which might have seemed like a clever idea in 1941, could not easily be used to incorporate the pattern of functionality characteristic of the actual hydroaromatic steroids. By 1950, when Woodward's first work on the dienone-phenol rearrangement appeared,[11] he was in hot pursuit of the synthesis of hydroaromatic steroids by other methods and was about to leap-frog the competition[25] by reaching that goal without having passed through the intermediate stage of the ring A aromatics.

4.5 A Mechanistic Motivation

Woodward's dienone-phenol rearrangement study[11] thus can be considered a natural offshoot of an abandoned approach to steroid total synthesis. However, it was far more than just an intellectual salvage effort. As we have seen, he had conjectured that the steroidal dienone rearrangement products were not the ring A aromatic 3-hydroxy compounds that others had supposed. Should this prove to be correct, he would have alerted the world that attempts to make estrogens by means of such rearrangements might prove fruitless. Beyond that, the mechanistic challenge of the rearrangements provided motivation in its own right. It came at a time of intense interest in carbonium ion rearrangements,[26] and Woodward must have felt (justifiably) that his immersion in dienone chemistry had given him advantages that could be applied toward the solution of such questions. In fact (see below), it is a reasonable surmise that the anticipation of a decisive *mechanistic* experiment in this area was the main motivation for all of Woodward's work on the dienone-phenol rearrangement.

He was eventually to carry out two experiments, the first[11] of which inspired a succession of studies by others with results which, although experimentally reliable, led to bewilderingly inconsistent mechanistic conclusions. The second Woodward experiment appeared to be a decisive contribution toward the solution of a subtle and deeply significant problem which lies at the heart of carbonium ion chemistry. We shall examine the actual significance of the the second experiment with the advantage of four decades of posteriority. But before we do so, it will be helpful to learn more about the dienone-phenol rearrangement, beginning with Woodward's first experiment.

4.5.1 Woodward's First Experiment. The Model Dienone

Scheme 6 shows the rearrangements of the steroidal substrates **10** and **12** as formulated by Inhoffen. Woodward recognized that direct proof of the structures of the products would be difficult and proposed to study the behavior of the *model* dienone 38 (Scheme 10). If the rearrangement of **38** were to take the methyl migration route analogous to the one Inhoffen proposed for **10** and **12** (Scheme 6), the product would be 5,6,7,8-tetrahydro-4-methyl-2-naphthol **39** (Scheme 10).

Scheme 10

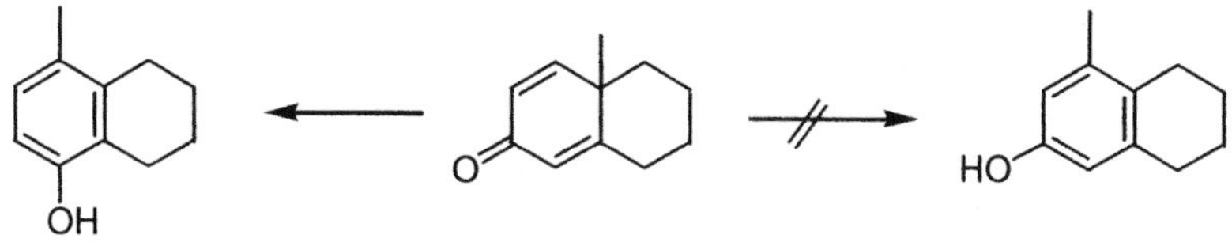

OH O HO

40 **38** **39**

However, this product **39** already was known and was synthesized again by Woodward and Singh[11] by unambiguous methods. The synthetic comparison compound proved to be different from the rearrangement product from **38**, which was thus shown not to possess structure **39**. Instead, its true structure was shown to be that of a product of *methylene* migration, namely 5,6,7,8-tetrahydro-4-methyl-1-naphthol **40**; confirmation of the structure was achieved by synthesis of an authentic sample of that substance and direct comparison. Apparently, the acid-catalyzed rearrangement of the model dienone **38**, under the same Ac_2O/H_2SO_4 conditions as Inhoffen had used (Scheme 6) for the rearrangements of **10** and **12**, takes a different course from the methyl migration observed in the rearrangement of santonin (Scheme 5).[8,27] Since the latter rearrangement had formed the analogical basis for Inhoffen's original assignments of structures **15** and **16** to the products from the rearrangements of the steroidal dienones (Scheme 6), the latter assignments now had to be reconsidered.

Woodward and Singh proposed that in the rearrangement of the model dienone **38**, the initially formed conjugate acid **38**-H^+ suffered migration of a ring member methylene group instead of methyl migration (Scheme 11).

They envisioned two possible detailed mechanisms for the overall rearrangement **38**-H^+ $\rightarrow$ **40:** either two successive 1,2-migrations passing over the spiro intermediate **41** (**38**-H^+ $\rightarrow$ **41** $\rightarrow$ **43** $\rightarrow$ **40**), or a reaction of **38**-H^+ in which "the migrating group accepts the positive charge from the ring, and the cationic center in the resulting intermediate (**42**) attacks the ring in the position ortho to the hydroxyl group in a normal electrophilic substitution reaction." Intermolecular and intramolecular cation-arene and cation-alkene "π-complexes" similar to the intermediate **42**, recently had been extensively discussed by Dewar[28] and although Woodward did not use this nomenclature, there is no doubt that he was aware of the idea.[29] In a later section (4.5.4) of this chapter, under the heading "Woodward's Second Experiment," we describe efforts to distinguish between the two subtly different mechanisms of Scheme 11, which for convenience we call a *direct 1,3-rearrangement* (**38**-H$^+$ → **42** → **43** → **40**) and a *pair of successive 1,2-rearrangements via a spiro intermediate* (**38**-H$^+$ → **41** → **43** → **40**). Whether the π-complex **42** in the direct 1,3-rearrangement pathway is a true metastable intermediate or a transition state is not specified.

Scheme 11

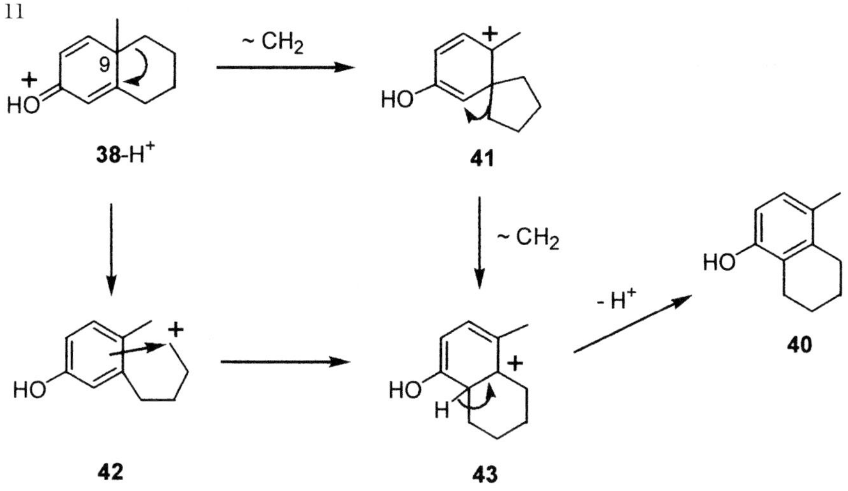

At this stage, aside from santonin (Scheme 5), which may be considered a special case,[27] there were two examples of dienones whose phenolic structures were firmly based on experimental evidence: the naphthalenoid system **17** of Wilds and Djerassi (Scheme 7), in which methyl migration predominates, and the model bicyclic dienone **38** of Woodward and Singh (Scheme 10), in which methylene migration predominates. Woodward and Singh[11] pointed out that the different results in the latter two instances did not really amount to a contradiction, since in the rearrangement of the Wilds-Djerassi[12] naphthalenoid dienone **17**, regardless of mechanistic details, methylene migration (Scheme 12) to give product **44**, for example, would require torsion about the bond (marked in boldface) connecting the 5-position of the dienone system with the naphthalene ring in the conjugate acid **17**-H$^+$. The reader may be best able to visualize this process by mentally trying to connect the two carbon atoms marked a and b in **17**-H$^+$, which must be achieved to effect the reaction **17**-H$^+$ → **44**. Since the boldface bond has partial double bond character, the necessary torsion would be energetically costly, the methylene migration pathway would be suppressed, and the competing methyl migration to give **18** would predominate (Schemes 7 and 12).

Woodward and Singh then made the daring proposal that the phenols obtained by Inhoffen from rearrangements of the steroidal dienones **10** and **12** did not have the structures **15** and **16** derived by methyl migration (Scheme 6), but instead were real-

Scheme 12

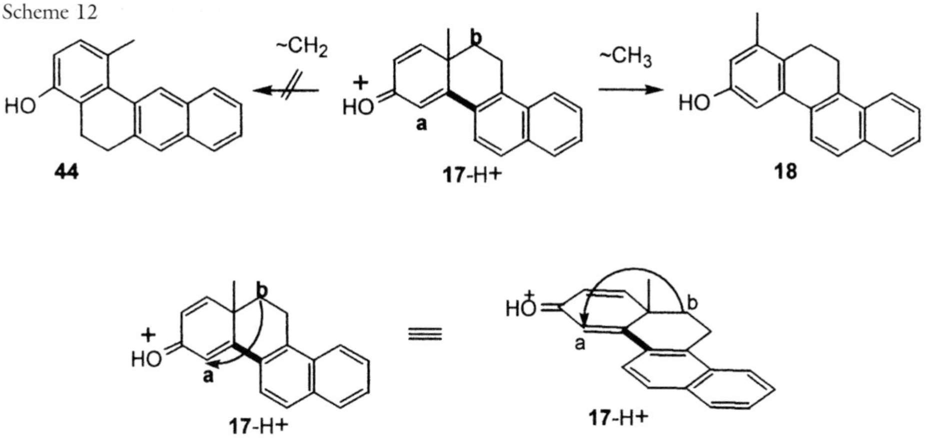

~CH2 ~CH3 HO a b 44 17-H+ 18

HO a b 17-H+ 17-H+

ly the products of ring member migration, either **45** or **46** (Scheme 13), with **46** preferred. Structure **46** would result from rearrangement of ring carbon C_8 from C_9 to C_4, either directly or through a spiro cationic intermediate (not shown here) anal-

Scheme 13

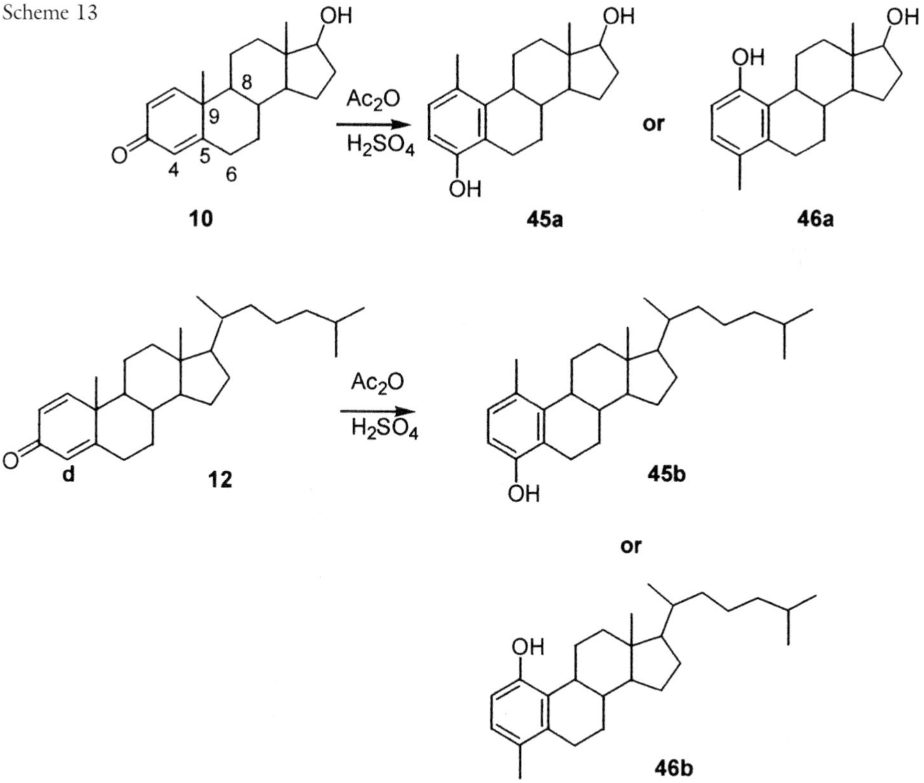

OH 8 9 5 4 6 Ac_2O H_2SO_4 10 45a or 46a

d 12 Ac_2O H_2SO_4 45b or 46b

ogous to 41 (Scheme 11). Structure 45 would result from migration of C_8 to C_5 and C_6 to C_4. He also suggested that similar corrections should be made of the structures of the phenolic products from other steroidal dienones whose rearrangements had been reported in the literature. These assignments were based entirely on analogy, since no additional evidence beyond the behavior of the model compound **38** (Scheme 10) was then available.

Soon after the appearance of the Woodward-Singh paper,[11] Djerassi and co-workers[30] reported a test of Woodward's hypothesis that methyl migration would be favored relative to methylene migration because of restricted rotation about the internuclear bond, as in the Wilds-Djerassi example[12] of Schemes 7 and 12. Thus, the same kind of restriction should apply in the case of the *trienedione*, $\Delta^{1,4,6}$-androstatrien-3, 17-dione **47**, which then should give the methyl migration product **48** (Scheme 14). The experiment[30] strongly suggested that this expectation was fulfilled. The phenolic rearrangement product upon catalytic hydrogenation gave 1-methylestrone **49**. In contrast to the steroidal rearrangement products obtained from the *dienone* system by Inhoffen (Scheme 6), this substance **49** and its dehydro precursor **48** were both soluble in aqueous alkali. Also, **49** was estrogenic with about half the potency of natural estrone. Thus, the two properties that were "missing" in the Inhoffen products and that had led Woodward to question Inhoffen's structural assignments, were present in the phenol **49**. Similar results were obtained[30] when the C_{17} alcohol derived by reduction of the carbonyl group of **47** was subjected to the same conditions. The assignments of the methyl-migrated structures shown in Scheme 14 were subsequently directly confirmed by Dreiding and Pummer.[31] They degraded the Djerassi ketone **49** to the tricyclic ether **50** and independently synthesized the latter by unambiguous methods.

Scheme 14

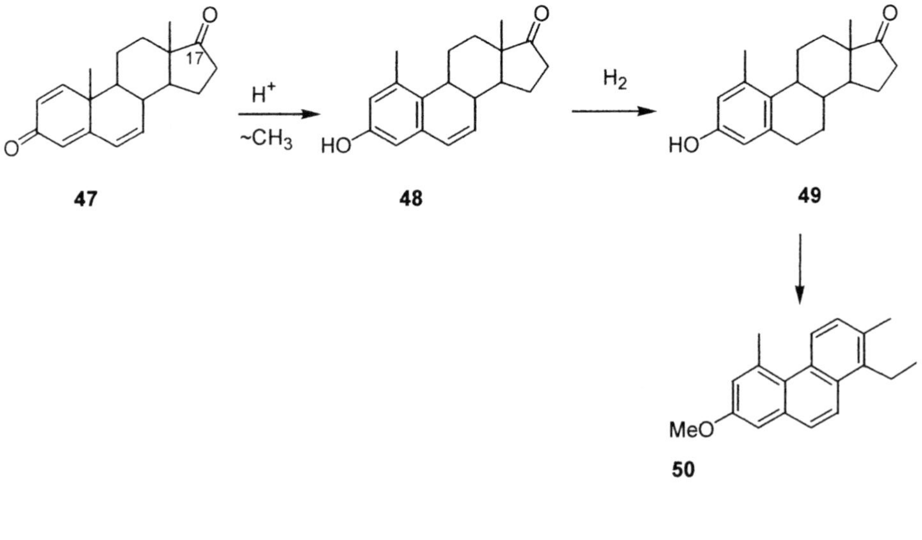

47 —(H^+, ~CH_3)→ 48 —(H_2)→ 49 → 50

Woodward and Singh's re-interpretation of cholestadienone rearrangement as **12** → **46b** (Scheme 13) was finally confirmed, by the synthesis[32] of **51**, the selenium dehydrogenation product of the phenol **46b** (Scheme 15).

Scheme 15

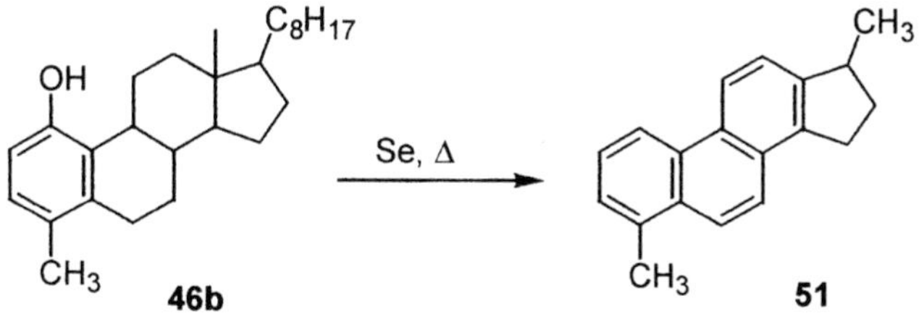

C_8H_{17} OH CH_3 **46b** Se, Δ CH_3 CH_3 **51**

Of course, all this was an impressive validation of Woodward's analysis and seemed to justify his assumption[11] that in the absence of what he called "special factors," such as the presence of a conjugating group that restricted internal rotation, the migration of the ring member observed in the dienone model compound (**38**, Schemes 10 and 11) would generally be preferred over migration of the bridgehead methyl group. Note that Woodward never offerred a *reason* for the observed preference but merely postulated the likelihood of uniform behavior in related compounds. Although there is no question that Woodward's structural assignments were correct, his apparent conviction that uniform behavior would prevail in the mechanistic analysis was soon to be proven illusory. As we shall now see, the course of the dienone-phenol rearrangement would be shown by others to be exquisitively sensitive to the nature of the reaction conditions and to apparently small changes in the structure of the substrate. In fact, these changes bring to light the existence of several different mechanisms.

4.5.2 Subtleties and Complications. The Effect of Reaction Conditions

The first inkling of the extraordinary variability of the dienone-phenol rearrangement came in an important 1953 paper of Dreiding and co-workers.[33a] For reasons not made clear in their publication, these authors investigated the effect of changing the nature of the acidic medium from the essentially anhydrous acetic anhydride-concentrated sulfuric acid conditions used in the prior steroid examples[33b] to aqueous hydrochloric or hydrobromic acid. They gave no precedent for expecting a change in outcome, and in fact they pointed out that in the case of santonin (Scheme 5), the only prior compound to which both anhydrous and aqueous rearrangement conditions had been applied, the same product (from methyl migration) was obtained. They found that, in contrast to the ring-member migration observed in prior steroid dienones under the anhydrous conditions (for example, see Scheme 13),[33a] the rearrangement of androstadienedione **52** in *aqueous* HCl or HBr led to *methyl* migration, giving 1-methylestrone **49** as the major product (Scheme 16).

Scheme 16

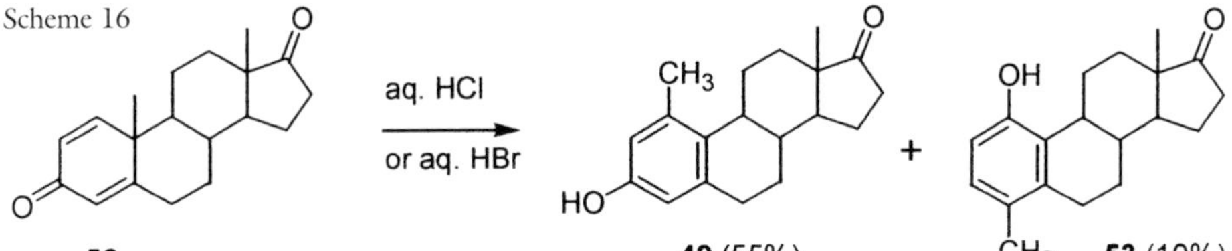

aq. HCl or aq. HBr

52 → **49** (55%) + **53** (10%)

In a careful re-examination of the reaction products, they took advantage of the solubility of **49** in alkali to separate it from the ring-member migration product **53** (Scheme 16), which is formed in minor amount and which is cryptophenolic.

This startling result called for a re-examination of the Woodward-Singh model dienone **38** (Scheme 10), which had been reported[11] to give only the methylene rearrangement product **40** by the *anhydrous* route. Dreiding[33] found (Scheme 17) that in *aqueous* mineral acid, the behavior of **38** switched over to give predominantly methyl migration product **39** instead. Only a small amount of **40** also was present, as determined by isolation from the alkali-insoluble cryptophenolic fraction of the reaction mixture. In a much later study in 1965, using the more accurate analytical method of gas chromatography,[34] Hopff and Dreiding showed that the methyl:methylene migration (**39:40**) product ratio from **38** changed from 20:80 in the Ac_2O route to 80:20 in aqueous acid (Scheme 17).

Scheme 17

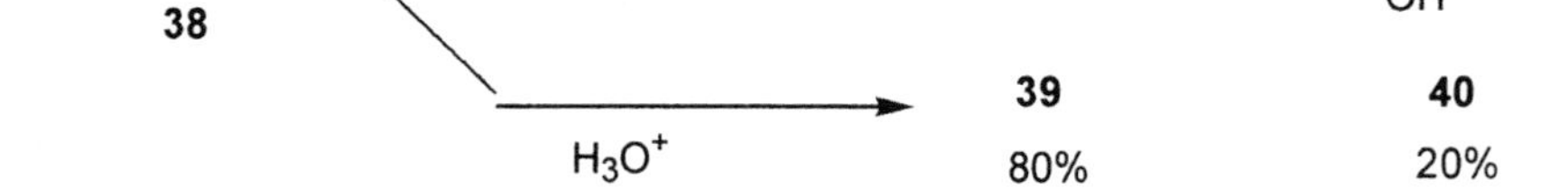

38: 1. Ac_2O, H_2SO4; 2. aq. KOH; 3. H_3O^+ → **39** 20% + **40** 80%

38: H_3O^+ → **39** 80% + **40** 20%

Woodward's earlier conclusion[11] that only the product of ring member migration **40** was formed in the original study in Ac_2O was based upon examination of the product isolated in 58% yield after crystallization from methanol. The small amount of methyl migration product **39** which was presumably present could have been detected had the deacetylated reaction mixture been extracted with alkali. Clearly, the outcomes of these dienone-phenol rearrangements demonstrate a competition between at least two closely matched pathways. It seems a reasonable conjecture that had Woodward found the minor product **39** in his work, he might have tempered his generalization of the preference for methylene migration.

4.5.3 Further Complications. Alternative Rearrangement Routes. The Bridgehead → Bridgehead Methyl Shift

In 1963, Caspi and Grover[35] raised the possibility that an alternative mechanism might be involved in the methyl shifts observed in the dienone-phenol rearrangements (see Scheme 18). The methyl shifts had been formulated as simple 1,2-migrations of the C-19 methyl from C-10 to C-1 of the steroid nucleus, but no experimental evidence ruled out a more complex pathway in which the methyl shift occurred initially from bridgehead to bridgehead (C-10 to C-5) and was followed by methylene migration. Such C-10 → C-5 methyl shifts had been observed in the Westphalen rearrangement (sometimes called the Westphalen-Lettré rearrangement). In the case of androstadienedione **52**, either the direct C-10 → C-1 methyl shift or the C-10 → C-5 methyl shift followed by methylene rearrangement would give the same product, 1-methylestrone **49**, as had been observed before (Scheme 16). However, the pathways would be distinguishable by an isotopic labeling experiment (Scheme 18). Experimentally, Caspi and Grover showed that rearrangement of labeled androstadienedione 52-*4-^{14}C* under aq. HCl/HOAc catalysis gave a mixture containing predominantly the 1-methyl-3-hydroxy aromatized product **49**, in which the ^{14}C label was exclusively at C-4. None of the label was found at C-2 of **49**, where it would have lodged if the rearrangement had gone by the C-10 → C-5 methyl shift/methylene shift sequence. Therefore, in the present instance, the simple methyl shift, from C-10 to C-1, is essentially the only operative pathway for the formation of **49**.

Scheme 18

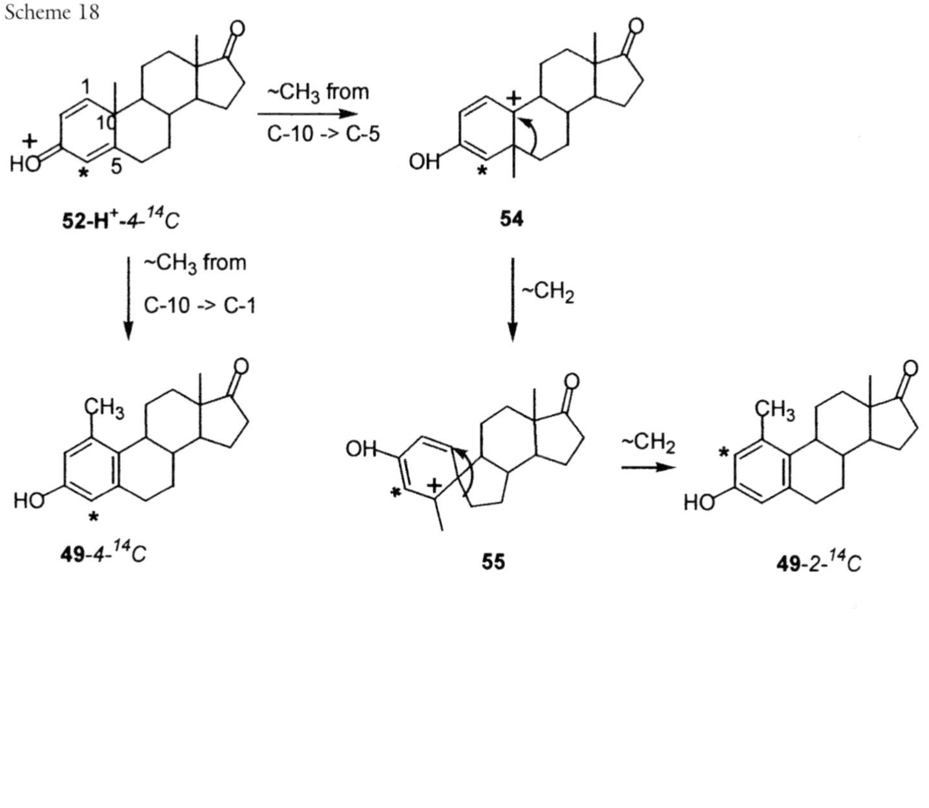

One might be tempted to conclude that Caspi and Grover had wasted their time investigating a far-fetched mechanistic hypothesis which involved an indirect and perversely complicated pathway. However, almost immediately after their paper appeared, Kropp[36] used a methyl-labeled dienone **56** to show that the idea had merit. For example, the reaction mixture from the aqueous mineral acid-catalyzed rearrangement of the dienone **56** (Scheme 19) contained not only the products of direct C-10 → C-1 methyl migration **57** and of methylene migration **58**, but also a product **59** apparently formed by the same type of successive Westphalen-then-methylene migration (see Scheme 18) that Caspi and Grover had sought in the steroidal system.

Scheme 19

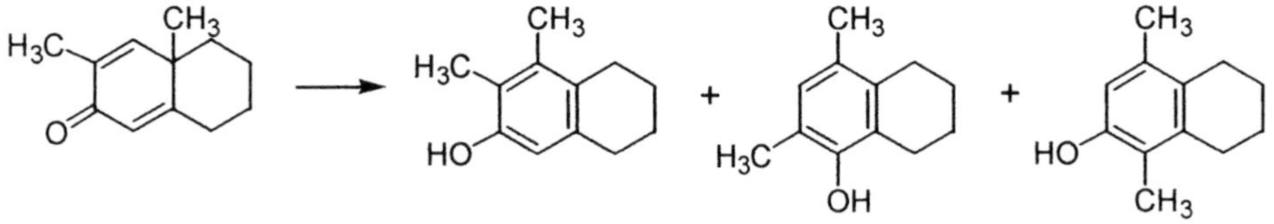

56 **57** **58** **59**

In fact, even in the rearrangement of the prototypical Woodward-Singh dienone **38** under the Dreiding aqueous mineral acid conditions, the major product, the tetralol **39** (Scheme 17), was subsequently shown, in a ^{14}C-labeling experiment by Futaki,[37] to result from the Westphalen-then-methylene migration mechanism. Apparently, a 1-methyl-3-hydroxy product from these dienone rearrangements is accessible either by a direct C-10 → C-1 methyl shift or by a Westphalen-then-methylene migration mechanism. Which pathway is actually followed in a particular case depends, in some intricate and still not understood way, on the structure of the substrate.

In the light of the knowledge gained from these experiments, an observer would have to conclude that Woodward was fortunate that his application of the ring member migration mechanism to Inhoffen's steroidal cases turned out to be correct. The later work, as we have seen, disclosed the great mechanistic variability of the dienone-phenol rearrangement and thereby conflicted with a basic assumption of Woodward's analysis, namely a uniform preference for ring member migration. It might be said, therefore, that his success in elucidating Inhoffen's rearrangements was a spectacular example of being right for the wrong reasons.

4.5.4 Woodward's Second Experiment. The Methylene Migrations: Successive 1,2-Rearrangements or 1,3-Rearrangement via a π-Complex?

Perhaps the most intensely studied problem in physical organic chemistry during the immediate post-World War II period was the mechanism of nucleophilic substitution at carbon. The questions included the stereochemistry of solvolytic nucleophilic displacements and the detailed nature of the cationic intermediates involved in solvolyses and in a wide variety of molecular rearrangements. At the International Colloquium held in Montpellier in 1950,[26] there was extensive discussion of the bridged ions (Scheme 20) thought to be responsible for the "neighboring group effect" and its characteristic outcome of stereochemical retention in substitution reactions. Winstein, whose experimental work played a decisive role in establishing the effect,[38] had formulated the intermediates (or transition states) as resonance hybrids,[39] using the superposition of canonical structures as advocated in the truncated valence bond (VB) theory of Pauling. Thus, the bromonium ion **60** and the bridged ions in carbonium ion rearrangements **61** were hybrids of the structures shown in Scheme 20. Subsequently, Dewar[28] suggested the π-complex notation **62** and **63** for alkene complexes and **64** for arene complexes. This was based upon the molecular orbital (MO) theory, which had first been applied to organic molecules by Hückel.[40]

Scheme 20

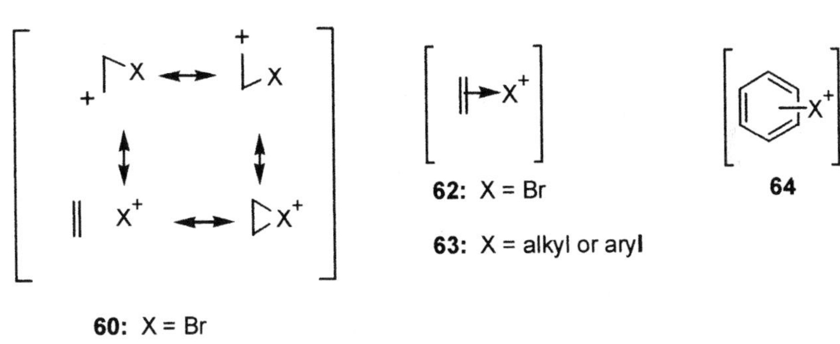

At the time, there was considerable skepticism that the π-complex hypothesis was a necessary or even a useful addition to the theory of organic cationic species.[41,42] One of the major objections came from Winstein,[41,43] who pointed out that the contending representations would be very difficult to distinguish structurally, since the cyclic resonance hybrid (**60** or **61**) and the π-complex (**62** or **63**) both had a characteristic triangular arrangement of the atoms. Dewar[28b,43] admitted that both representations were triangular (hence in a sense cyclic) but emphasized one advantage of the π-complex notation: it predicts that cationic π-complexes should be more stable than anionic ones, which is in agreemnet with the observation that such anionic complexes generally do not occur. Thus, the MO description for a cationic π-complex represents the interaction of the filled π-orbital of the alkene with the low-lying empty orbital (LUMO) of the X^+ species. In an anionic π-complex, the lone-pair electrons of the X^- species already occupy the corresponding orbital, and the electrons to be donated by the alkene therefore must go into much higher energy orbitals.

On the other hand, as Dewar noted, the resonance notation has an unsatisfactory feature: since exactly the same number and kind of contributing resonance structures can be written with an anionic as with a cationic bridged ion, one would expect that such anionic species also should exist, but apparently they did not. The failure of the Pauling VB theory here corresponds, of course, to its similar difficulty (see Chapter 3) in accounting for aromatic and antiaromatic properties as a function of electron count in cyclic conjugated molecules, for example in the electronically analogous case of cyclopropenium cation (stable aromatic) vs cyclopropenide anion (unstable antiaromatic).[40] These debates had great significance in that period, when theoretical chemistry had not yet developed quantitative computational approaches to molecular structure, and chemists were forced to rely on qualitative models derived from MO and VB theory. Dewar deserves a good share of the credit for calling attention to the utility of MO-derived models.

In retrospect, I believe that much of the early resistance to the π-complex idea arose from Dewar's attempt to make a spectacular demonstration of it by resolving the tangled literature on the rearrangement of hydrazobenzene to benzidine. In a single bold stroke,[28a] he rationalized the extensive prior investigations with a mechanistic proposal that featured an arene π-complex intermediate of the type **64** (Scheme 20). Unfortunately, he was insufficiently precise in specifying the kinetic behavior to be expected, and readers were entirely justified in the assumption that for his mechanism as written, the reaction should be first-order in hydrazobenzene and first-order in $[H^+]$. Soon after his proposal appeared,[28a] the reaction was shown[44] to be *second-order* in $[H^+]$ under some circumstances. When confronted with this result at the Montpellier conference,[42] Dewar[45] gave a not wholly implausible explanation of how his mechanism could fit the observed kinetics without abandoning the π-complex intermediate. However, as the *cognoscenti* present at the conference probably realized, the new kinetic findings were just the tip of the iceberg. It was to be years before the extremely complex kinetic pattern of these rearrangements was explored,[46] but the reluctance to jump on to the π-complex bandwagon in 1950 is understandable. It is ironic that the acceptance of the general idea of π-complexes, which later turned out to be fruitful,[46] was delayed by an early questionable application to a complicated example. For an analogous bump in the road to full cognitive definition of carbonium ion rearrangements, see the discussion of the non-vicinal hydrogen shifts in bicyclic cations (Section 5.9).

In any case, it is clear that in the period around 1950 (and for many years thereafter), the question of carbonium ion structure was prominent. An intriguing aspect of the debate was that the protagonists on both sides seemed to be unable to design experimental tests for the intervention of π-complexes in carbonium ion rearrangements.

Woodward and Singh[11] recognized their opportunity to resolve this problem:

» *Differentiation between the possible mechanisms [that is, the direct 1,3-rearrangement viaa* π*-complex intermediate or transition state versus the successive 1,2-rearrangements via a spiro cation, as exemplified in Scheme 11] is of considerable importance in connection with the the detailed mechanism of Wagner-Meerwein rearrangements in general, and the matter deserves clarification through the study of appropriately substituted dienones; such experiments are now under way in our Laboratory.* «[11]

Although the details of of this "second experiment" were never published,[47–49] Woodward did give the overall conclusion in an essay he contributed to a multi-author volume honoring Sir Robert Robinson.[47] The substrate **65**-H^+ (Schemes 21 and 22) incorporates the dienone system of the first Woodward-Singh experiment[11] (Scheme 11) but now contains a third ring cis-fused to the cyclohexane moiety. If **65**-H^+ containing one enantiomer in excess were subjected to the rearrangement conditions, the two contending mechanisms would give very different results. A 1,3-rearrangement through a π-complex **66** (Scheme 21) would retain the chirality of the starting material, so that an excess of one enantiomer, **68a**, of the phenolic product would be formed.

Scheme 21

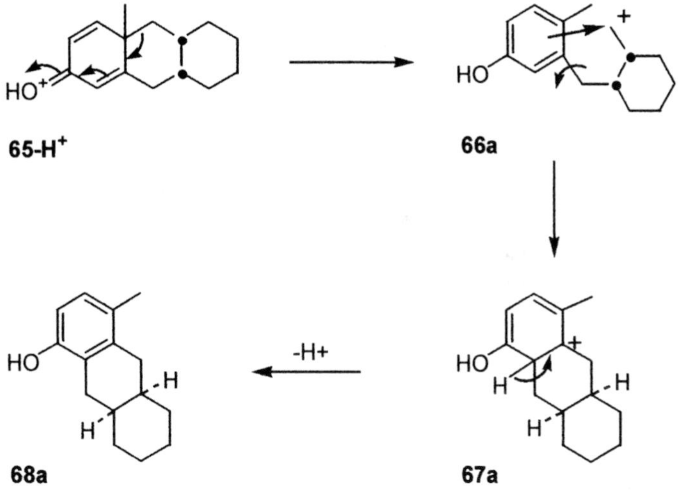

On the other hand, successive 1, 2-rearrangements (Scheme 22) by way of the conventional carbonium ions **69** and **67a–67b** would lead to racemic product (equal amounts of **68a** and **68b**), since the spiro structure of **69** would place the OH-bearing ring in a plane of symmetry bisecting the perhydroindane moiety, and migrations of carbons a and b then would become equally probable.[47]

Note that a similar experiment could not be carried out with nonracemic dienone **38** in the original Woodward-Singh system (Scheme 11). Although the π-complex intermediate **42** might be formed in a chiral condition, it could enantiomerize by rel-

Scheme 22

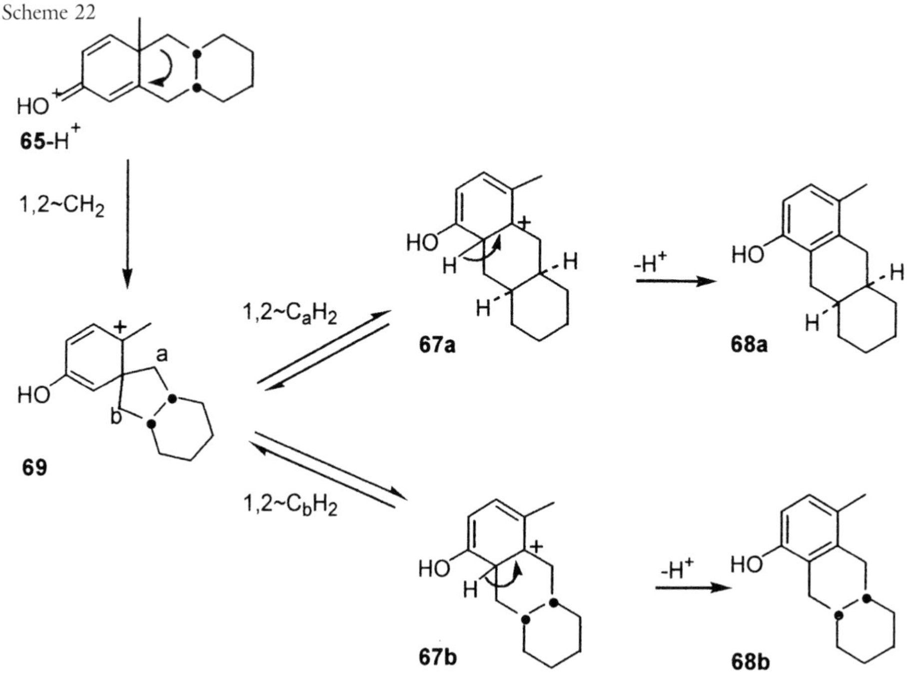

ative internal rotations of the arene ring and the side chain. Thus, the formation of racemic product need not be attributed to the formation of the achiral spiro intermediate **41**. On the other hand, the experiment of Schemes 21–22 is cleverly designed to provide a π-complex (**66a**, Scheme 21) whose chirality remains intact even if similar internal rotations should occur, since the stereo centers at the ring fusion positions are not perturbed by this process. Racemic product can only be formed if the meso spiro intermediate (**69**, Scheme 22) is on the pathway.

Here is Woodward's laconic account[47] of the outcome of this ingenious experiment:

» *Our examination of the rearrangement of the optically active ketone (***65***) revealed that the change led to a racemic product. Clearly, the molecule must pass into a symmetrical condition; the only symmetrical intermediate which can be envisaged is (***69***), and the change is a succession of 1,2-rearrangements rather than a 1,3-rearrangement involving a generalized intermediate.* «

Fulfillment of the requirement for meso character of the spiro intermediate, of course necessitates a cis, not a trans, fusion of the decalin ring system of the starting dienone **65**. Beyond that, however, there is a more subtle dynamic issue: There is no guarantee that the spiro intermediate **69** will be born with a true plane of symmetry. Some conformational adjustments may be needed to produce effective meso character. Unless these are fast on the time scale of the second rearrangement step, carbons a and b may not have a chance to become equivalent. This kind of "memory effect" has been studied in a number of multiple rearrangements of carbonium ions;[50] the general problem of symmetrization experiments and symmetrical intermediates is discussed in Chapter 5 of this book and in references given there.

Subsequently, Bloom[51] demonstrated the intervention of a spiro intermediate in the rearrangement of another dienone, and Futaki[52] later reported a confirmatory result which showed the intervention of a spiro intermediate in the Ac_2O/H_2SO_4-induced rearrangement of the isotopically labeled form, **38**-*9^{14}C*, of the original Woodward-Singh dienone (Scheme 11) to the acetate of the phenol **40**, which showed an approximately equal distribution of the label at the two α-methylene positions.

The Woodward "second experiment,"[47] the Bloom experiment,[51] and the Futaki experiment[52] each leave no doubt that a spiro intermediate is accessible in these rearrangements. Bloom stated his conclusion with appropriate caution, stating that his result "demonstrates the intermediacy of the spiran intermediate in the reaction studied."[51] As cited above, however, Woodward's conclusion was subtly but significantly different: "the change is a succession of 1,2-rearrangements [through a spiro intermediate] *rather than* a 1,3-rearrangement involving a generalized intermediate,"[47] *i.e.*, the π-complex **66a** (Scheme 21) [emphasis supplied here]. If the words "rather than" are supposed to suggest that there is a mechanistic preference for a faster reaction through the spiro intermediate than through the π-complex, the interpretation is permitted but not required by the experimental facts presented. It would be possible to imagine, for example, that the 1,3-rearrangement via the π-complex actually occurs faster than the 1,2-rearrangement via the spiro intermediate (see Scheme 21), but that the rearranged intermediate **67a** from that pathway (Scheme 21) is converted to the spiro cation by a reversal (Scheme 22, **67a** → **69**) of the second step of the 1,2-rearrangement. If the latter step is fast compared to the deprotonation step, racemized phenolic product would also be observed. For this reason, it is very difficult to decide experimentally which rearrangement pathway is *kinetically* preferred. All one can say is that the spiro cation is accessible, either because it is formed faster than the π-complex or because the cationic intermediate from the direct 1,3-rearrangement (via a π-complex) reverts to the spiro cation faster than it deprotonates.[53]

4.5.5 The Direct 1,3-Rearrangement Mechanism in the Phenol-Phenol Rearrangement

Thus, the two tests for the direct mechanism just described each failed to find definitive evidence for or against it as an accessible pathway. The existence of the direct mechanism was first disclosed in careful isotopic labeling experiments by Futaki,[54] who studied the so-called "phenol-phenol rearrangement" discovered by Hopff and Dreiding.[34] Although neither the acetic anhydride/conc. sulfuric acid conditions nor the 30–50% aqueous mineral acid conditions previously employed in the dienone-phenol rearrangements causes secondary rearrangements of the phenols,[34,36] such reactions can be observed under somewhat more vigorous conditions (70% perchloric acid, 80° C).[54,8] For example, Hopff and Dreiding[34] found that the tetralol **39** (Scheme 23) gave a rearranged isomer **72** upon such treatment. They described the rearrangement as protonation of the aromatic ring at the bridgehead position para to the hydroxyl group, followed by successive 1,2-shifts via a spiro cation **71** (Scheme 23).

Scheme 23

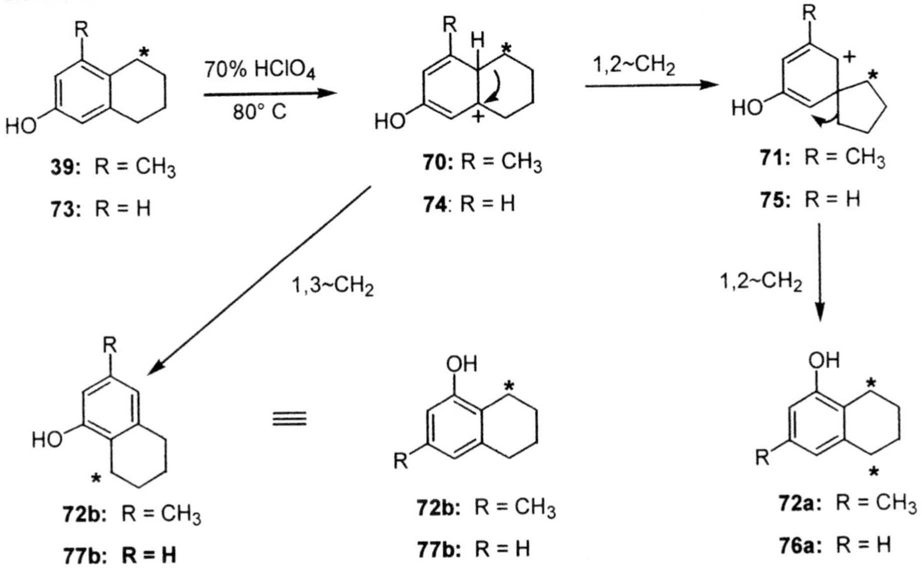

70% $HClO_4$, 80° C

1,2~CH_2

1,3~CH_2

1,2~CH_2

39: R = CH_3
73: R = H

70: R = CH_3
74: R = H

71: R = CH_3
75: R = H

72b: R = CH_3
77b: R = H

72b: R = CH_3
77b: R = H

72a: R = CH_3
76a: R = H

However, Futaki showed[54] that when the rearrangement was carried out on **39**-*^{14}C* labeled in the asterisked position (Scheme 23), the label appeared in the rearranged phenol nearly exclusively (98%) in the position shown in **72b**, rather than distributed equally between two positions, as in the product **72a** expected to result from the spiro intermediate **71**. *This result rules out any significant contribution of the spiro intermediate 1,2-shift mechanism and is entirely consistent with a dominant 1,3-shift pathway for the rearrangement in this case.* Moreover, he observed a similar result in the rearrangement of the unmethylated 2-tetralol **73**, which gives 1-tetralol **77b** with the label 96% specifically in the position shown.

4.6 Conclusions

The structural guiding principle of the dienone-phenol rearrangements apparently is that there is no guiding principle. All conceivable mechanisms seem to be available in a closely balanced array, and which of them dominates seems to be a delicate function of substrate structure and reaction conditions. Woodward's attempt to bring the steroid rearrangements under intellectual control by appeal to the model reaction of a synthetic dienone, as in the "first experiment" of Scheme 11, now seems almost naively optimistic. Yet, to judge yesterday's actions and thoughts by today's standards is a pernicious historiographical error. Fifty years ago, little was understood about these extraordinarily complex rearrangements. Even today, a full comprehension still eludes us. Although later workers, bit by bit, are shouldering open the stubborn door to the secret chamber of the dienone-phenol mysteries, we should remember that it was Woodward who unlocked it.

I several times heard Woodward, in his later years, state the dictum:[1b] "there are no general reactions." Usually, this was a warning issued in the context of a multistep synthesis of a complex natural product, to remind a co-worker, testing a new reaction on a simplified derivative, that even a successful outcome in the model system did not guarantee success in the actual application. Nevertheless, one may conjecture that these surprises early in his career, in the *mechanistic* study of the dienone-phenol rearrangement, may have been the experiences that started him toward his provocative formulation.

If Woodward is right, does this mean that organic chemistry is a completely unpredictable enterprise? Of course not. "There are no general reactions" does not mean "there are no *characteristic reactions*." Behavior patterns do carry over from one molecule to the next. Chemists recognize this in the concept of a *functional group*, a set of atoms that, regardless of which molecule they happen to be in, have an inherent tendency to react with a particular reagent in a particular way. Other parts of the molecule have only a low or negligible tendency to react with the reagent, so they are carried through to the product without change. However, there may well be still other portions of the molecule in which the tendency to react can be small or large. Thus, one must keep in mind that in making use of the infinite variability of organic chemical structure and reaction conditions, the chemist may encounter some instances in which reaction at the targeted site becomes a minor pathway, and side-reactions at the alternative potentially vulnerable units become dominant. A major task then is to find means to guide the outcome along the desired channel, to make the reaction, in a word, selective. Of course, in that endeavor, Woodward was an unsurpassed master. But that is another story.

4.7 Acknowledgments

I thank Professors H.H. Wasserman, P.J. Kropp, and R. Hoffmann for information, helpful comments, and reminiscences. Mr. D.A. Ware and other members of the staff of the Harvard University Archives made available relevant files from their collection of documents of R.B. Woodward.

4.8 References

(1) (a) For a list of reviews of the dienone-phenol rearrangement, see March, J. *Advanced Organic Chemistry*, 4th ed. Wiley Interscience, New York, NY, 1992, p. 1079. (b) The epigraph can be found in: Woodward, R.B. *Proc. Robert A. Welch Foundation Conference on Chemical Research, XII. Organic Synthesis*, **1969**, 3.

(2) (a) Windaus was soon after to win a Nobel Prize (in 1928) for his work on Vitamin D, cholesterol, and related steroids. Although by 1932, it was clear that Windaus's structural formula for the steroid nucleus was incorrect, the significance of his pioneering contributions fully justified the award. An insightful and sympathetic account of these events is given in ref. 2b, p. 71ff. (b) Fieser, L.F.; Fieser, M. Steroids, Reinhold, New York, 1959, Chapter 15.

(3) In the structural formulas of this chapter, stereochemical configurations are specified only when they are needed for reference to the text.

(4) (a) Inhoffen, H.H.; Zühlsdorf, G. *Ber.* **1941**, *74*, 1911. (b) Windaus's generosity in turning over problems to his former co-workers was admirable but apparently not unique. For example, Otto Wallach of Göttingen was awarded the Nobel Prize in Chemistry in 1910 "for his services to organic chemistry and the chemical industry by his pioneer work in the field of alicyclic compounds." Much of this work concerned the terpenes (see Chapter 5) and originated with a group of samples of essential oils given to him by his mentor, August Kekulé. Kekulé thought obtaining fruitful or important results from research on such intractable mixtures of natural products to be too difficult and hence not suitable to advance Wallach's career. Ultimately, Wallach succeeded in persuading Kekulé to relinquish these materials, which he had stored untouched for fifteen years. (see ref.4c) (c) Asimov, I. in *Asimov's Biographical Encyclopedia of Science and Technology*, 2nd. ed. Doubleday, New York, 1982, p. 790.

(5) Review: Ref. 2b, p. 104ff.

(6) The pyrolysis process was subsequently improved by others and eventually came into commercial production: Review: Ref. 2b, pp. 479–480.

(7) Inhoffen, H.H.; Huang-Minlon. *Naturwiss.* **1938**, *26*, 756.

(8) Clemo,G.R.; Haworth, R.D.; Walton, E. *J. Chem. Soc.* **1930**, 1110.

(9) Inhoffen, H.H.; Zühlsdorf, G. *Ber.* **1941**, *74*, 604.

(10) For a list of references, see ref. 11.

(11) Woodward, R.B.; Singh, T. *J. Am. Chem. Soc.* **1950,** *72*, 494.

(12) Wilds, A.L.; Djerassi, C. *J. Am. Chem. Soc.* **1946,** *68*, 1712, 1716.

(13) Review and references: Ref. 2b, p. 460ff.

(14) (a) Bachmann, W.E.; Cole, W.; Wilds, A.L. *J. Am. Chem. Soc.* **1939,** *61*, 974. (b) *J. Am. Chem. Soc.* **1940,** *62*, 824.

(15) See the discussion of synthetic efficiency by: Ho, T.L. *Tactics of Organic Synthesis,* Wiley-Interscience, New York, NY, 1994, Chapter 1.

(16) Woodward, R.B. *J. Am. Chem. Soc.* **1940,** *62*, 1478, 1625.

(17) Woodward, R.B. Ph. D. Thesis, "A Synthetic Attack on the Oestrone Problem," Massachusetts Institute of Technology, 1937, typed draft copy, Woodward Papers, HUG (FP) 68.6, Series A, Box 10, Harvard University Archives.

(18) Barton, D.H.R.; Wasserman, H.H. Obituary of R.B. Woodward, *Tetrahedron*, **1979,** *35*, No. 19.

(19) Wasserman, H.H. Recollections conveyed in personal communication with the author, June 4, 1997.

(20) Woodward, R.B. Handwritten "yellow sheet" notes, of a conversation with Wasserman, H.H., Fall, 1941, in Private Papers, Wasserman, H.H. Cited with permission of the owner.
(21) Bachmann, W.E.; Kushner, S.; Stevenson, A.C. *J. Am. Chem. Soc.* **1942,** *64*, 974.
(22) Anner, G.; Miescher, K. *Helv. Chim. Acta*, **1948,** *31*, 2173 and subsequent papers.
(23) Johnson, W.S.; Banerjee, D.K.; Schneider, W.P.; Gutsche, C.D. *J. Am. Chem. Soc.* **1950,** *72*, 1426, and later papers from the Johnson group.
(24) (a) Woodward, R.B.; Sondheimer, F.; Taub, D.; Heusler, K.; McLamore, W.M. *J. Am. Chem. Soc.* **1951,** *73*, 2403. (b) Woodward, R.B.; Sondheimer, F.; Taub, D.; *ibid.* **1951,** *73*, 4057. (c) Woodward, R.B.; Sondheimer, F.; Taub, D. *J. Am. Chem. Soc.* **1952,** *74*, 4223.
(25) (a) He did not quite succeed in leap-frogging *all* of the competition. A communication, reporting the total synthesis of racemic methyl 3-keto-$\Delta^{4,9(11),16}$-etiocholatrienate,[24a] the "first synthesis of a compound possessing the full hydroaromatic steroid nucleus with the correct (natural) stereochemical configuration," contains the priority-claiming footnote: "First announced at the Centenary Lecture of the Chemical Society presented at Burlington House, London, on April 26, 1951." Almost simultaneously that same spring, Robinson and co-workers[25b] reported the total synthesis of another hydroaromatic steroid androstan-3β-ol-17-one (epiandrosterone). In a later paper,[24c] Woodward apparently tried to reinforce his claim of priority by repeating the date of his April 26 lecture and pointing out that the date of the issue of *Chemistry and Industry* in which the Robinson communication appeared was May 19. Although Chemistry and Industry did not indicate the dates of receipt of communications, it is my opinion that a fair-minded estimate of the delay between receipt and actual publication leaves no doubt that Robinson's work was complete at about the same time as Woodward's, or close enough to completion to render pursuit of the priority issue of interest only to extreme partisans. We may be sure, however, that it was of intense concern to the protagonists. (b) Cardwell, H.M.E.; Cornforth, J.W.; Duff, S.R.; Holterman, H.; Robinson, R. *Chemistry and Industry*, **1951,** 389.
(26) See for example the special issue: *Bull. Soc. Chim. France* **1951,** 1Cff., a compilation of papers and discussions presented at the *Colloque International sur les réarrangements moléculaires et l'inversion de Walden*, held in Montpellier, April 24–29, 1950.
(27) In the case of santonin (**13,** Scheme 5), as Woodward pointed out,[11] the methyl group at C_4 would disfavor methylene migration.
(28) (a) Dewar, M.J.S. *The Electronic Theory of Organic Chemistry*, Oxford, 1949, pp. 233–240. (b) Dewar, M.J.S. *Bull. Soc. Chim. France* **1951,** 71C.
(29) (a) Although the published versions of the Woodward-Singh paper of 1950 and the Woodward 1956 paper avoid the term "π-complex," an earlier handwritten draft[29b] of the 1950 paper uses it freely and with undisguised skepticism. Woodward's forbearance in muting his criticism for publication is admirable, but there can be little doubt that he intended the 1956 experiment to be an unequivocal refutation of the concept, at least in the dienone-phenol cases, and by extension, elsewhere. (b) Woodward Papers, HUG 68.8, Box 25, "Papers T. Singh," Harvard University Archives.
(30) Djerassi, C.; Rosenkranz, G.; Romo, J.; Pataki, J.; Kaufmann, St. *J. Am. Chem. Soc.* **1950,** *72*, 4540.
(31) Dreiding, A.S.; Pummer, W.J. *J. Am. Chem. Soc.* **1953,** *75*, 3162.
(32) Woodward, R.B.; Inhoffen, H.H.; Larson, H.O.; Menzel, K.-H. *Ber.* **1953,** *86*, 594.

(33) (a) Dreiding, A.S.; Pummer, W.J.; Tomasewski, A.J. *J. Am. Chem. Soc.* **1953,** *75*, 3159. (b) References given in ref. 33a.
(34) Hopff, W.H.; Dreiding, A.S. *Angew. Chem. Intl. Ed. Engl.* **1965,** *4*, 690.
(35) Caspi, E.; Grover, P.K. *Tetrahedron Lett.* **1963,** 591.
(36) Kropp, P.J. *Tetrahedron Lett.* **1963,** 1671.
(37) Futaki, R. *Tetrahedron Lett.* **1967,** 2455.
(38) Review: Bartlett, P.D. *J. Am. Chem. Soc.* **1972,** *94*, 2161.
(39) Winstein, S. *Bull. Soc. Chim. France* **1951,** 53C.
(40) For a review of the early history of the VB and MO methods, see: Berson, J.A. *Angew. Chem. Intl. Ed. Engl.* **1996,** *35*, 2750. See also Chapter 3 of this book.
(41) Winstein, S. Question in the discussion at the Montpellier conference following Dewar's lecture (ref. 28b), *Bull. Soc. Chim. France* **1951,** 78C.
(42) Bartlett, P.D. Question in the discussion at the Montpellier conference following Dewar's lecture (ref. 28b), *Bull. Soc. Chim. France* **1951,** 78C.
(43) Not far below the surface of Winstein's objection[41] one can discern the motivation to protect priority. In this instance, the issue goes beyond mere territoriality. After all, there is value in questioning whether a new hypothesis deserves to replace an existing one unless the new one can be shown to have distinct advantages and to be meaningfully different. At a conference in Cleveland in 1968, Winstein asked a similar question following a presentation by Olah on "corner protonated cylopropanes," namely, how did they differ from carbon-bridged ions? Although in describing the relationship of π-complexes and bridged ions, Dewar in fact admitted[45] "the present interpretation of their structure (as π-complexes) only differs in terminology and is not radically new," his explanation of the differences in the MO and VB origins of the two descriptions seems to me to be conceptually substantial.
(44) Hammond, G.S.; Shine, H.J. *J. Am. Chem. Soc.* **1950,** *72*, 220.
(45) Dewar, M.J.S Answer to the question of ref. 42, *Bull. Soc. Chim. France* **1951,** 78C.
(46) For references to the long history of controversy over the nature and role of π-complexes, see: March, J. *Advanced Organic Chemistry*, 4th ed., Wiley-Interscience, New York, NY, 1992, pp. 79–82, 505–507.
(47) Woodward, R.B. in *Perspectives in Organic Chemistry*, Todd A.R., ed. Interscience, New York, NY, 1956, p. 178. No documentation was ever given in the open literature.
(48) In ref. 47, Woodward does not cite the names of collaborators on the "second experiment." Singh was a participant in the project, but it is not clear that he ever brought it to a conclusion. In fact, his Ph. D. Thesis,[49] describes how "it was decided to prepare this dienone (*i.e.*, **65**, Scheme 21) in optically active form and study its rearrangement to decide between the two possible mechanisms of the rearrangement reaction... *The work has not yet been completed.*" [emphasis supplied]. Whether he stayed on to finish the project after obtaining his degree, or whether one or more additional collaborators should be credited, I have so far been unable to determine.
(49) Singh, T. "Synthesis and Rearrangement of Cyclohexadienones," Ph. D. Thesis, Harvard University, 1950, p. 49 ff.
(50) (a) Berson, J.A. *Angew. Chem. Intl. Ed. Engl.* **1968,** *7*, 779; *Angew. Chem.* **1968,** *80*, 765. (b) Berson, J.A.; Gajewski, J.J; Donald, D.S. *J. Am. Chem. Soc.* **1969,** *91*, 5550. (c) Berson, J.A.; Poonian, M.S.; Libbey, W.J. *J. Am. Chem. Soc.* **1969,** *91*, 5567. (d) Berson, J.A.; Donald, D.S.; Libbey, W.J. *J. Am. Chem. Soc.* **1969,** *91*, 5580. (e) Berson, J.A.; Wege, D.; Clarke, G.M.; Bergman, R.G. *J. Am. Chem. Soc.* **1969,** *91*, 5594. (f) Berson, J.A.; Bergman, R.G.; Clarke, G.M.; Wege, D. *J. Am. Chem. Soc.* **1969,** *91*,

5601. (g) Berson, J.A.; McKenna, J.M.; Junge, H. *J. Am. Chem. Soc.* **1971,** *93*, 1296. (h) Berson, J.A.; Foley, J.W. *J. Am. Chem. Soc.* **1971,** *93*, 1297. (i) Berson, J.A.; Luibrand, R.T.; Kundu, N.G.; Morris, D.G. *J. Am. Chem. Soc.* **1971,** *93*, 3075.

(51) (a) Bloom, S.M. *J. Am. Chem. Soc.* **1958,** *80*, 6280. (b) Bloom, S.M. *J. Org. Chem.* **1959,** *24*, 278.

(52) Futaki, R. *Tetrahedron Lett.* **1964,** 3059.

(53) On the other hand, once actual phenolic product is formed, achievement of a conventional thermodynamic equilibrium and consequent racemization by reversion of the *phenol* back through the cationic intermediates to its enantiomer (**68a** → **67a** → **69** → **67b** → **68b**), a phenol-phenol rearrangement, seems unlikely under these conditions (refs. 34, 36).

(54) Futaki, R. *Tetrahedron Lett.* **1968,** 6245.

Chapter 5

Meditations on the Special Convictive Power of Symmetrization Experiments

5.1 Introduction

I derived the following generalizations introspectively, but subsequent informal polling of colleagues confirmed my guess that they are widely accepted in the community of chemists:

The superstructure of interconnected assumptions and beliefs about our discipline that we accept as true, or at least useful, is held together by a few definitive, persuasive, and illuminating experiments. Among such experiments, those which involve what I shall call "reaction symmetrization" hold a place of special regard, even of delight and affection.

In this chapter, through several illustrative cases, I attempt to comprehend the extraordinary impact of symmetrization experiments, a decisive power that seems to contain not only intellectual but also esthetic, emotional, and even spiritual components.

The kind of symmetrization I consider here is not the one in which an unsymmetrical reactant gives a symmetrical product. That transformation usually has only trivial mechanistic or synthetic significance. For example, the hydrogenation of propene, an unsymmetrical molecule, gives propane, a symmetrical one, whereas the semihydrogenation of 1,3-butadiene, a symmetrical molecule, gives 1-butene, an unsymmetrical one. The distinction between these two reactions is purely accidental, and I would argue that the two are extremely similar.

A far more telling experiment is one that results in "reaction symmetrization." For the present purpose, it is useful to adopt this term to refer to formation of equal amounts of two enantiomeric or two or more symmetry-related isotopomeric products. One large category of such reaction symmetrizations includes isotopic exchange reactions between an atom in the reactant and an isotopic atom, either in the external environment (isotope incorporation or depletion) or elsewhere in the reactant (re-location of isotopic position). A second category includes racemizations by interconversions of enantiomers. This operational classification is based upon the experimental procedures used, but in a broader sense, the categories have in common the idea that some element of stereochemical or isotopic positional specificity originally present in the reactant, potentially or actually, can be randomized or scrambled in the product. The unique property of processes occurring with "reaction symmetrization" is that, as we shall see, the *reaction pathways* are symmetrical. This is the justification for using the term for the reactions under our definition, even though either the reactant, or the product, or both may be unsymmetrical. We recognize a conceptual relationship of such symmetrization reactions to *quasi-symmetrizations*, such as catalyzed epimerizations, in which two diastereomers interconvert. These latter do not incorporate reaction symmetry in our sense; epimerizations scramble the configuration at one site but maintain it at the non-reacting stereogenic center(s) throughout the reaction. Nevertheless, they are analogous to true symmetrizations in that they too result in the formation of products in which stereochemical or structural specificity is lost.

Aside from the difference in the structural sense between symmetrization and quasi-symmetrization, there is another interesting formal distinction. When the randomization occurs by interconversion of the reactant with its own stereoisomer or isotopomer, the thermodynamic change associated with the achievement of equilibrium includes a positive entropy of mixing. Because the entropy change on mixing two substances is greatest when they are present in equal amount,[1] the entropy of mixing, Rln 2 (about 1.4 cal/mol/degree), of racemization or isotopic position exchange (symmetrization by our definition), which necessarily leads to equal amounts of two species, will be greater than that of diastereomerization (quasi-symmetrization by our definition), which in general does not.

We have used the term "equal amounts of two species" to refer to the result of racemization or isotopic exchange, but of course, exact symmetrization may not be achieved experimentally in isotopic exchange experiments, even if the intermediate would be symmetrical were no isotopic label affixed, since the mechanistic "equivalences" in general are perturbed by kinetic or equilibrium isotope effects. Even when the experimental design uses racemization as the criterion, a curious quirk of nature, the "electro-weak advantage"[2] associated with parity non-conservation, is expected to produce a deviation of about one part in 10^{17} from strict equality of the amounts of enantiomeric products. Because the deviation is many orders of magnitude too small to be detected by our currently crude analytical ability, it has no practical significance in most chemical problems. Nevertheless, one is bemused by its philosophical implications. Why is the universe constructed forever to deny us (by this tiny bit!) the attainment of perfection? Moreover, why should we have to bear the knowledge of this proscription?

Whatever the answers to these (perhaps imponderable) questions, the use of symmetry and quasi-symmetry criteria has a long history in mechanistic investigation. This chapter searches out the beginnings of the ideas in some specific early cases and traces later events in the sometimes stumbling progress to understanding.

5.2 Enolization as a Mechanism of Symmetrization. The Menthone Problem

As far as I have been able to determine, the first ascription of a symmetrization (actually, quasi-symmetrization) phenomenon to a symmetrical (actually quasi-symmetrical) intermediate occurred in 1889 in Beckmann's study[3] of the so-called "inversion" of menthone. This ketone originally was obtained by the oxidation of menthol, the familiar odoriferous principle of mint oils, but subsequently was found as such in nature.[4] Beckmann's work on it gives another example of the motivating force of natural products chemistry on mechanistic research.[5] Although serious errors marred his seminal 1889 paper, the idea survived that stereochemical specificity could be lost as a consequence of reaction through an unstable intermediate, in this case, an enol.

At that time, Beckmann did not question the then accepted structure **1** (Scheme 1) for menthone. Later work[4,6] showed that the C_3H_7 group is isopropyl instead of propyl and that the positions of the methyl and C_3H_7 groups must be interchanged, as in structure **2** (shown here in the stereochemical configuration eventually established as that of the most abundant natural isomer). Actually, these changes had no effect on the main thrust of the argument but may have contributed to the apparent reluctance of some later authors to acknowledge Beckmann's insight that enolization and stereomutation are intimately connected.

Scheme 1

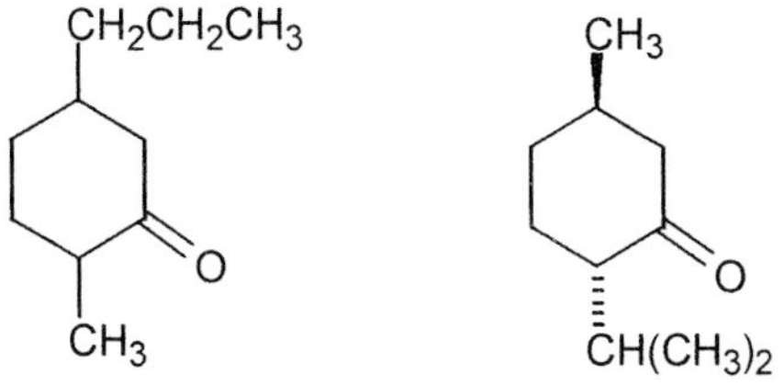

$CH_2CH_2CH_3$ / O / CH_3

1

CH_3 / O / $CH(CH_3)_2$

2

Beckmann found that dissolution of an optically active levorotatory sample of menthone ("Linksmenthon"), $[\alpha]_D$-28.18°, in concentrated sulfuric acid followed by quenching of the reaction mixture on ice gave a new ketonic material ("Rechtsmenthon") $[\alpha]_D$ +26.33° to +28.14°. Similar "inversions" of rotation to varying degrees were observed in alkaline media. Beckmann recognized that "the transformations belong to the previously puzzling group of phenomena which Berzelius has called 'catalytic.'"

Although the rotations of the two ketones were opposite in sign and nearly equal in magnitude, Beckmann seems to have realized that they were not enantiomers, since they gave oximes of different melting points and different optical rotations. However, his strangely convoluted and in a sense, self-contradictory, discussion revealed much confusion about the stereochemical issues. These misunderstandings are perhaps best presented by direct quotation:

» *In the sense of LeBel and van't Hoff, one asymmetric carbon atom is present in menthone (3), namely the one marked with*. The four different atoms or atomic groupings bound to it can be thought of as lying at the corners of an irregular tetrahedron (4), as follows (Scheme 2).*

Scheme 2

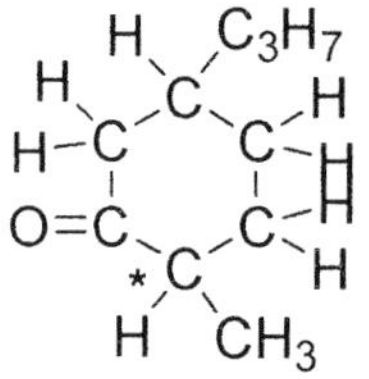

3 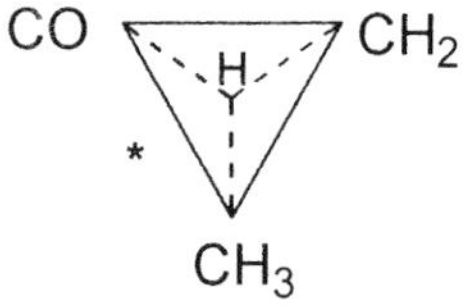 **4**

This system requires a certain optical rotation, for example, levo. If however the CO and H, for example, exchange places or if any single exchange of that kind occurs, the tetrahedron changes immediately into the mirror-image tetrahedron. The result will be the reversal of the original levo rotation into a dextro rotation of equal magnitude. «

Today, any student of elementary organic chemistry would recognize after a semester in the course that menthone contains two asymmetric carbon atoms, and that whether Beckmann's hypothetical exchange at one center actually would result in a change of the sign of rotation could not be readily predicted but would depend on the contribution of the second center. Moreover, in an actual experiment, the final observed rotation at equilibrium would not be equal and opposite to the starting rotation but rather would be the algebraic sum of the rotations of the two diastereisomers, appropriately weighted by their respective concentrations in the resulting mixture.

We must remember, however, that the LeBel-van't Hoff theory was only fifteen years old in 1889, so that many chemists still were unfamiliar with all of its implications. Also, as we shall see in Section 5.6, even decades later, some major figures in organic chemistry still had not mastered the elementary consequences of the theory.

In Beckmann's defense, we can say that he actually had a more nearly correct idea of the true stereochemical situation than he was able to express clearly. In other words, a present-day student who answered this problem on a class test as Beckmann did would receive a grade of "partial credit".

Beckmann demonstrated the nature of his idea of enolization-induced stereomutation **3a** → **3b** → **3c** (Scheme 3) with van t'Hoff-LeBel tetrahedra **5** → **6** → **7** (Scheme 4).

Scheme 3

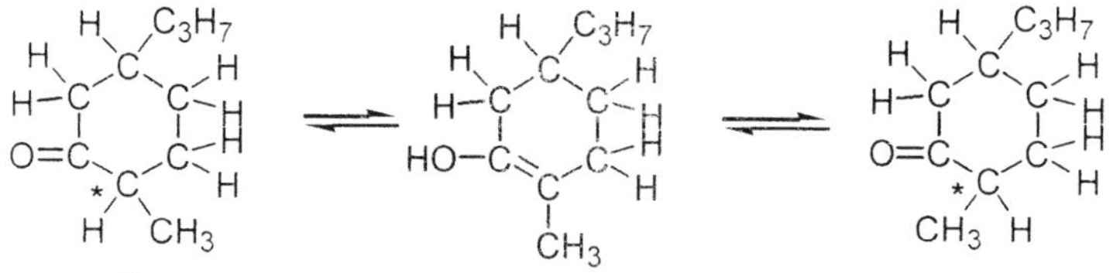

3a ⇌ **3b** ⇌ **3c**

Scheme 4

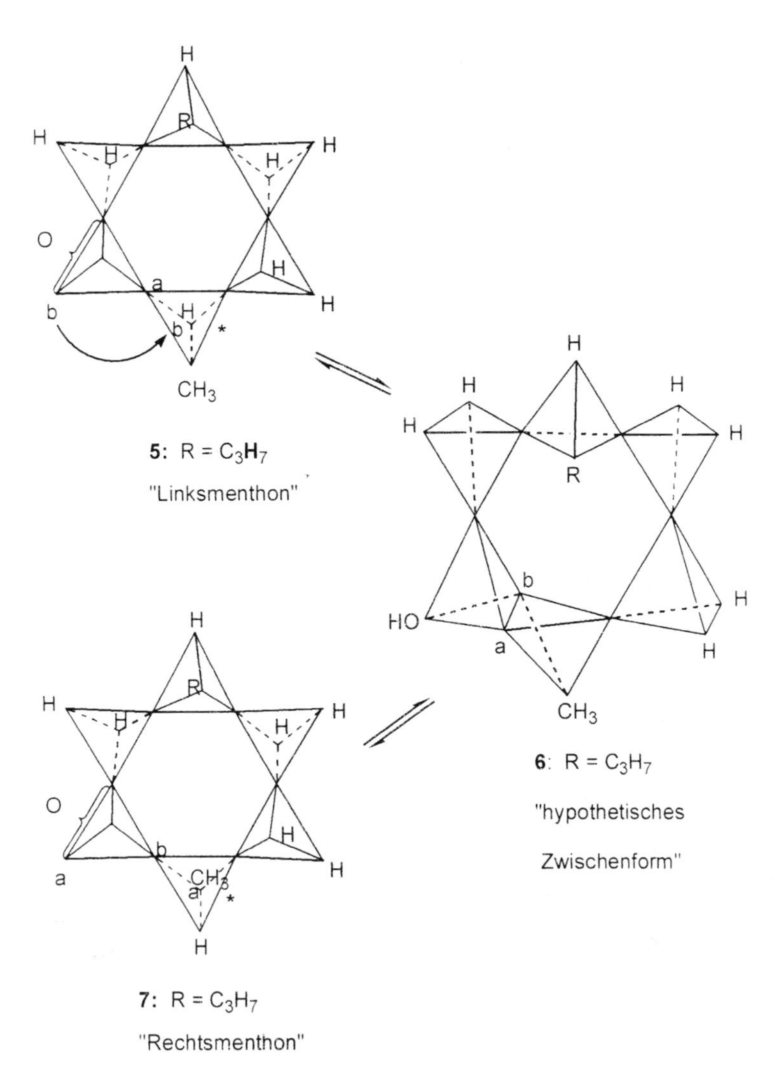

The asymmetric carbon he identified in the starting menthone **3a** (**5**) and in the stereomutated menthone **3c** (7) is marked with an asterisk. He postulated enolization in the words:

» *One may assume that it is probable that the atomic grouping OC-CHCH3 permits the transient existence of the tautomeric form HOC=CCH3. The presence of acids and bases, that is, compounds that can react with hydroxyl, will only favor that result.* «

He visualized the intermediate enol form as **6** and recognized that reconversion to the keto form could occur by transferring the enolic hydrogen either to either of the stero-defining sites a or b on the doubly bonded methyl-bearing enolic carbon, thus generating either the starting ketone **5** or the stereoisomeric ketone 7.

Although Beckmann gave no references to support this hypothesis, the idea of tautomerism, and of enols as reactive forms of carbonyl compounds, was certainly prominent in the literature at that time, thanks especially to the work of Laar, Baeyer, Michael, and Butlerov.[7] We may assume therefore that the structural change of enolization was an accepted part of chemists' knowledge in 1889. What was new was Beckmann's application of this to the stereomutation problem.

We should note that Beckmann apparently did not have a complete grasp of the use of the tetrahedral representations in Scheme 4, since the drawing of **6** seems to indicate that (inexplicably) an inversion of the configuration at the C_3H_7-bearing carbon has occurred. It is deeply instructive to compare Beckmann's clumsy tetrahedron notation, derived from van't Hoff and LeBel, to the streamlined modern representation (Scheme 5, **8–10**) of the same transformations. We can sometimes lose sight of just how important to organic chemists is a succinct graphical way of showing three-dimensional information in two dimensions. An unanswered question is whether better notation led to greater stereochemical understanding or vice versa. In any case, the early chemists often stumbled over stereochemical issues we take for granted today.

Scheme 5

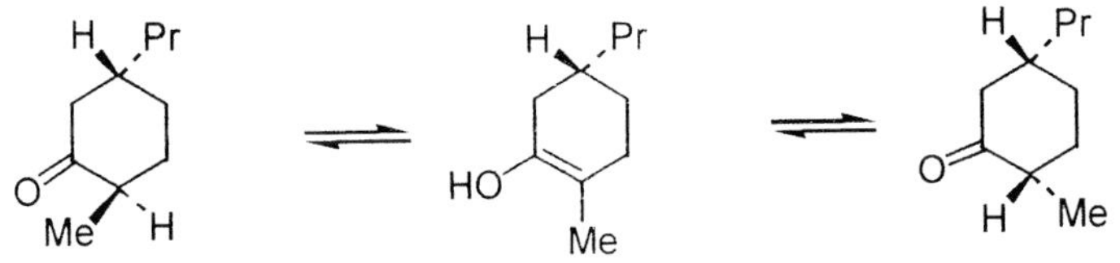

H Pr / O / Me H — **8** ⇌ H Pr / HO / Me — **9** ⇌ H Pr / O / H Me — **10**

Although Beckmann was clearly incorrect in thinking that menthone has only one asymmetric center, he nevertheless understood that the relationship between "Linksmenthon" and "Rechtsmenthon" is one of cis-trans isomerism, and he seems to have grasped why one could not expect to obtain equal quantities of the two in the ketonization of their common enol:

» *Although the methyl group affects both sides of the double bond equally, the bond positions a and b are obviously unequally influenced by the (substituent at) the para position [see* 6*, Scheme 4]. Position b is on the side of a hydrogen atom and position a on the side of a propyl group. Thus it is to be expected that the terminal carbon atom of the latter can directly approach position a, whose opening leads to "Rechtsmenthon."* «

Unfortunately, we cannot completely absolve him, since he immediately afterwards offerred another perplexing and internally contradictory conclusion:

» *That the final products of inversion [that is, stereomutation] under all conditions, even if often only to a moderate degree, are found to be dextrorotatory, can be explained as a greater stability of the dextrorotatory compound.* «

Of course, this statement would be necessarily true only if the optically rotatory powers of the pure dextro and levo materials were equal and opposite, which in general would not be the case unless the isomers were enantiomers. But if they were enantiomers, the rotation at equilibrium would be zero, not finite and positive.

Seven years passed before Beckmann returned to the menthone problem,[8] a period during which he was busy with other projects, including the ramifications of the important discovery[9] of what came to be called the Beckmann rearrangement:[10]

» *In the year 1887 [actually 1889], I published an article under the above title which was concerned with the mutual relationship of menthone and menthol as well as the properties of these compounds in comparison with those of camphor and borneol.*

Phosphorus pentachloride treatment of the oxime obtained from menthone and hydroxylamine soon led to an isomer with entirely unexpected properties. For clarification of this, experiments with other oximes were carried out. These provided information on the rearrangement reaction that Victor Meyer kindly named after me and also led to stereoisomeric oximes. The wish to determine the molecular weights of the rearrangement products gave motivation for our laboratory to study Raoult's freezing point method and to enhance it by the development of the boiling point method. While these activities diverted me from the clarifications of the initial substances for a long time, other researchers showed that menthol has a constitution other than that which had been assumed previously. For that reason, the earlier conclusions now must be re-examined with the new foundations. I would like to be permitted to go into this more closely. «

In the meantime, in 1893 Wallach[11] had pointed out the errors in Beckmann's 1889 discussion[3] of the menthone "inversion" and had given a clear analysis of the stereochemical problems associated with two dissimilar asymmetric centers. Although much further work on the enolization mechanism of stereomutation, both mutarotation of diastereomers and racemizations of enantiomers, began to appear soon after, key omissions of appropriate citations resulted in the slighting of Beckmann's contributions. Lowry in 1899[12] studied the mutarotation of 3-nitrocamphor, and in rapid succession, Lapworth and Hann[13–15] studied the mutarotation of optically active esters of β-keto and β-aldehydo acids (for example, menthyl acetoacetate) via the enols, and Kipping and Hunter[16] attempted to racemize 2-benzylpropionic acid via the enol. None of these papers contain references to either Beckmann or Wallach. In 1904, Tutin and Kipping[17] discussed at some length the stereomutation of menthone, repeating almost verbatim (but without acknowledgment[18]) the same analysis that Wallach[11] had given eleven years before. They cited Beckmann's paper as reporting the "inversion" of menthone but did not point out that it was he who had recognized it to be a consequence of enolization. Justice finally triumphed after 46 years when Wagner-Jauregg's review[19] clearly credited this insight to Beckmann.

One can reasonably conjecture that a cause of the failures to make appropriate citation of Beckmann's work was that his credibility as a serious advocate of his own ideas was questionable. Beckmann's 1896 paper gave a discussion much improved over that of his 1889 paper, but failed to acknowledge Wallach's analysis. Moreover, in the 1896 paper, Beckmann mired himself into deeper difficulty with some simply incorrect speculations about the possibility of epimerization at the methyl-bearing stereogenic center during the acid- or based-catalyzed stereomutation. This center cannot become enolic, and Beckmann offerred no specific proposal on how the imagined epimerization could occur there. I have the impression that his confidence in the enolization mechanism was severely shaken when he realized the several errors of his earlier paper, and he may have decided that rather than persistence in what might prove to be indefensible, equivocation was the safer policy. This raises the issue of how much credit should be assigned an author who backs away from a proposal which subsequently is widely accepted by others. Here I believe that one must distinguish between the effect on the personal reputation of the author and the actual scientific utility of the idea itself. I have no doubt that the latter objective is the more worthy. A relevant and more graceful comment on this kind of problem is given by Willstätter:[20]

» *From our contributions to knowledge, permeated with errors, and from the contributions of our successors, who improve our work, the mistakes and mistaken teachings will vanish in the stream of self-purifying investigation, and our valid discoveries will go, even if for the most part anonymously, into the enduring legacy of humankind.* «

5.3 Tricyclene and the Wagner-Meerwein Rearrangement

In the examples of racemization and diastereomerization just discussed, we have seen how experimental evidence was used to detect the intervention of symmetrical (or quasi-symmetrical) intermediates. A major intellectual advance took place decades later, in which the *absence* of complete symmetrization was used to *exclude* a symmetrical species as an obligatory mechanistic feature. As we shall see, the controversy in this case revealed just how shallowly rooted were the fundamental principles of stereochemistry in the minds of some important chemists, even forty years after van't Hoff and LeBel.

In the period 1896–1900, Wagner discovered the ring rearrangement reaction that occurs when borneol and isoborneol, reduction products of the naturally occurring ketone camphor, are dehydrated to camphene (Scheme 6).

The history of those developments, which really caused a revolution in chemists' thinking, has been told elsewhere[21,22] and need not be repeated here in detail. In reading that literature, however, the modern chemist cannot escape a feeling of awe over the achievement of the early terpene workers in solving the structures of these intricately related materials at a time when the major physical instruments were balances and thermometers, the techniques of separation, purification, and identification were primitive, and even the fundamental intellectual structure of the subject was new and untested. Their publications leave no doubt that the terpenists knew how inadequate were the resources available to them. Despite such limitations, they pressed forward, combining plenty of the organic chemist's legendary *Gefühl*[23] with

Scheme 6

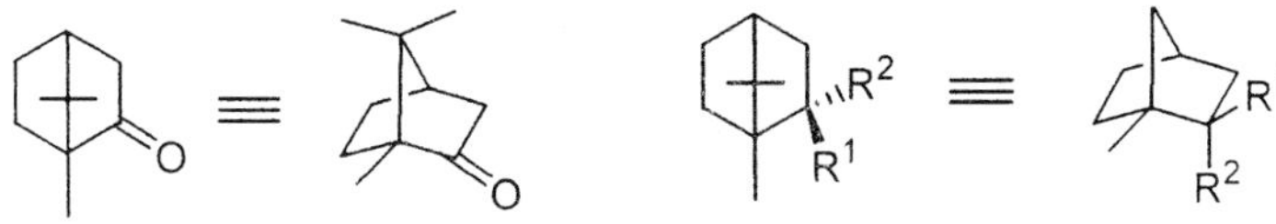

11, (+)-camphor

12, borneol: R^1 = H, R^2 = OH

13, isoborneol: R^1 = OH, R^2 = H

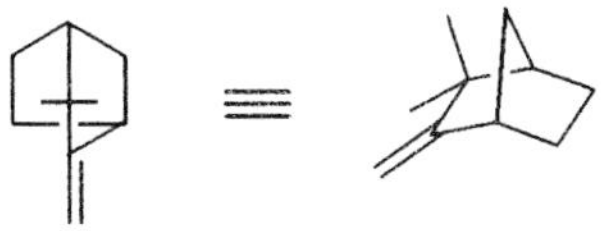

14, camphene

a certain valor and dash. At least one chemist, as a young student fifty years ago, looked upon those pioneers as intrepid heroes – and still does.

Wagner[24] recognized that the ring rearrangements he had demonstrated were closely related to the numerous pinacol (**15** → **16**) and retropinacol (**17** → **18**) rearrangements (shown in generalized form in Scheme 7) then already known. He was noncommittal on the mechanism of these transformations, but there already was under way a lively discussion in the literature on this problem, which dated back to the discovery of the pinacolic rearrangements by Fittig in 1860.[25]

Scheme 7

15, pinacol → **16, pinacolone**

17, pinacolyl alcohol → **18, tetraalkylethylene**

The chemist principally responsible for the crucial insights and experiments that led ultimately to a mechanistic understanding of the terpene rearrangements was Hans Meerwein (Figure 1).[26] By 1910, when he was 31 years old, he had become involved in the debates swirling around this issue. His publicly stated motivation[27] was couched in the customary cool, bloodless prose of the chemist:

» *It is known that the pinacol rearrangement and related rearrangement phenomena play an important role in terpene chemistry. The course is more complicated here than in the*

Figure 1. Hans Meerwein (1879–1965), Professor of Organic Chemistry, University of Marburg. Reproduced from ref. 26 with permission of Wiley-VCH publishers. Photo by Tita Binz, Mannheim.

aliphatic series, because the rearrangement occurs within the ring and is sometimes accompanied by remarkable alterations of the entire ring system.

*Because the complex structure of the terpenes and camphors renders these rearrangement phenomena obscure, it appeared to be of interest to study the course of the pinacol rearrangement in a simple cyclic case. To this end, we have prepared a pinacol (**19**) which, like the usual pinacols, contains on one hand two methyl groups but in which nevertheless the other two methyl groups are replaced by two methylene groups of a cyclopentane ring* « (see **19**).

CH_3
$C{-}CH_3$
OH OH

19

Yet I suspect that Meerwein's attraction to the field, like that of the student we met above, was driven at least in part by his admiration for the terpene pioneers and by his hunch that the implications of these rearrangements for all of chemistry transcended their significance in the field of natural products. We get some inkling of this in his 1914 paper:[28]

» *In the year 1899, in the same paper in which he proposed for the first time the formula for camphene that is generally accepted today, G. Wagner,[24] with remarkable insight, pointed out the parallelism between the formation of camphene from borneol and the rearrangement of pinacolyl alcohol to tetramethylethylene.* «

And later:

» *One will only be able to speak of the mechanism of the rearrangement of borneol to camphene if it is possible to elucidate the mechanism of the rearrangement that occurs in the transformation of pinacolyl alcohol to tetramethylethylene.* «

A lot of reading and some amateur psychologizing have persuaded me that the concept of reaction symmetrization played a crucial part in Meerwein's thinking as he developed his ideas about these rearrangements. In order to bring the reader to a position of agreement (or disagreement!) with me on this issue, I shall have to digress briefly to re-create the theoretical climate of that era.

5.4 The Pinacol Controversy

The focus of the problem evolved through time. Early investigators were concerned with whether the rearrangements involved stable intermediates, such as cyclopropanes[29] or epoxides,[30] or whether the true intermediates were unstable, transient species. A typical mechanism via a cyclopropane intermediate is shown in Scheme 8, **20–22**.

Scheme 8

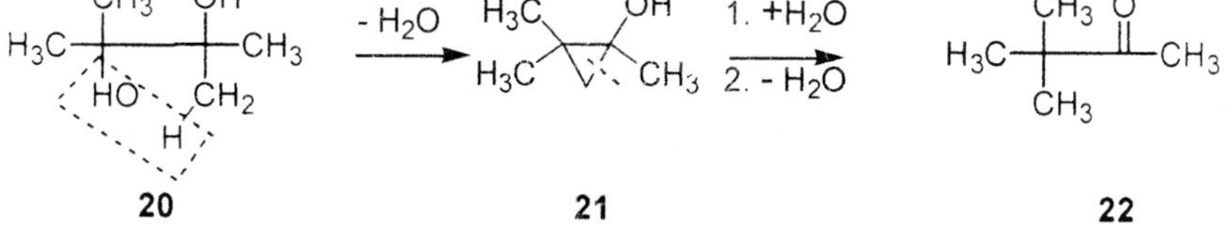

$\xrightarrow{-H_2O}$ $\xrightarrow[2.\ -H_2O]{1.\ +H_2O}$

20 **21** **22**

Later, the intermediacy of either cyclopropanes or epoxides was rendered highly improbable or actually impossible in several cases by straightforward experiments of several authors,[28,31,32,25b] which showed, for example, that the hypothetical intermediates reacted more slowly or gave different products than the substrates themselves. By exclusion, attention began to be directed toward transient species as the true intermediates. This pushed the debate onto a more sophisticated level: If these transients were in fact so unstable that they could not be isolated, how could one deduce their nature and demonstrate their existence?

We can enter the scene in the year 1907, when Tiffeneau published the text of a lecture he had presented at Haller's laboratory of organic chemistry at the Sorbonne in Paris.[32] This laid out a broad perspective of mechanistic proposals for a whole range of molecular rearrangements. In Tiffeneau's mind, what all these mechanisms had in common was the fact that they took place via unstable species, which now would be called *reactive intermediates*, and which did not adhere to the the standard valence numbers of stable compounds.

In the important case of the retropinacol rearrangement, for example, Tiffeneau proposed a process (**23** → **24** → **25,** Scheme 9) of carbon migration initiated by "liberation of two valences (⁓) on the same element," or in today's nomenclature, "α-elimination."

Scheme 9

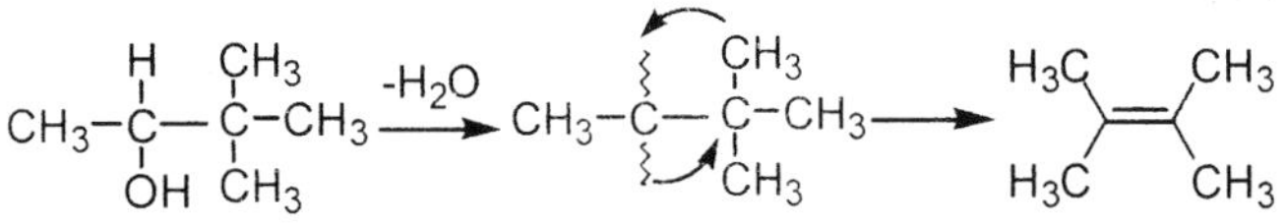

23 **24** **25**

To set this suggestion in the context of the time, we must note that the intermediate divalent carbon species **24** is related to similar "methylene" entities that had been proposed earlier by Nef[33] as intermediates in many reactions (which however, did not include pinacol rearrangements). Nef's ideas were greeted skeptically and refuted easily in several cases by many authors, particularly Michael.[34] The course of research often brings forth a basically false proposal that is nevertheless fruitful in the approach to the solution of a problem, a result characterized as being "right for the wrong reason." This sardonic but not entirely dismissive aphorism, however, does not really pertain to the suggestions of Nef. It is true that for some reactions, divalent neutral carbon species (carbenes) are now well accepted intermediates. For example, among the reactions for which Nef proposed methylene intermediates was the conversion of 1,1-dihaloethanes to terminal alkenes by reaction with metallic sodium, a transformation that probably does involve α-elimination to a carbene or carbenoid intermediate. However, such intermediates are not formed and play no role in most of the other processes for which he first proposed them. Thus, in the broad non-specific sense, Nef's proposal that methylenes could exist as transient species was not false, but his specific applications of it usually led nowhere.

Perhaps one of Nef's major contributions was in his influence on Stieglitz, his colleague at Chicago, who followed a productive path toward mechanistic understanding of rearrangements to nitrogen by making use of the similar (but in his case, correctly applied) concept of univalent neutral nitrogen intermediates (nitrenes). Stieglitz cited Nef's divalent carbon work in an early paper,[35a] but actually a more pertinent and prior origin of the univalent nitrogen idea is to be found in a truly insightful paper by Tiemann in 1891,[36] to which Stieglitz later referred.[35b] Tiemann's concise explanation of the Lossen rearrangement embodies α-elimination, which has proven to be a central hypothesis of the mechanisms of all the nitrogen rearrangements (Lossen, Curtius, Schmidt, Hofmann, etc.) and still plays that role more than a century later:

» *If one subjects benzohydroxamic acid $C_6H_5CONHOH$ to dry distillation, one obtains carbon dioxide and aniline. The decomposition presumably follows the course that water is split out and the residue, $C_6H_5CO.N:$, rearranges itself to $C_6H_5.N:CO$, that is carbanil (phenyl isocyanate), which with water at higher temperature dissociates to carbon dioxide and aniline.* «

5.5 Meerwein's First Hypothesis: Rearrangement via Divalent Carbon Intermediates

Thus, α-elimination was by no means an unfamiliar concept in 1907 when Tiffeneau attempted to apply it to the pinacol rearrangement. He gave no experimental evidence directly supporting its relevance to the task at hand, nor indeed could he have, since, as we now know, the idea is false in this application. Rather than support for divalent carbon intermediates, his reasoning stressed opposition to the idea of stable intermediates. Nevertheless, Tiffeneau's divalent carbon hypothesis for the pinacol rearrangement formed the basis for Meerwein's first proposal,[28] which was essentially an adoption of Tiffeneau's idea:

» *It appears to me that Tiffeneau has come close to the truth in his view that elimination of water from the same carbon atom is followed by migration of a radical with readjustment of the valences.* «

Meerwein's theoretical position thus was very similar to that of Tiffeneau, but his objective was more specifically the terpene rearrangements, an area in which he had a deep experimental knowledge. Like Tiffeneau, he rejected the idea of stable intermediates and was convinced that transient unstable species with unsaturated carbon valences must be involved. The actual experimental work of Meerwein's 1914 paper really offerred no compelling basis for the proposal of divalent carbon intermediates. Nevertheless, he took the opportunity of this publication to put forward arguments based largely upon critical analysis of the literature.

Although much of that discussion is inconclusive to a modern reader, there was one crucial point which takes on great significance in the context of our discussion of symmetrization criteria of mechanism. In the literature, application of the cyclopropane mechanism to the Wagner rearrangements had led to the hypothesis that the key intermediate in the prototypical terpene rearrangement, borneol → camphene, is *tricyclene* (**26**, Scheme 10), which could be imagined to be formed by γ-elimination of water from borneol **12**. Re-addition of water at another bond (C_6–C_1) of the newly formed cyclopropane ring, followed by dehydration of the intermediate camphene hydrate **27**, would lead to camphene **14**.

Meerwein proposed instead the mechanism shown in Scheme 11, which simply follows Tiffeneau's steps for the α-elimination in the acyclic case, with additional steps

Scheme 10

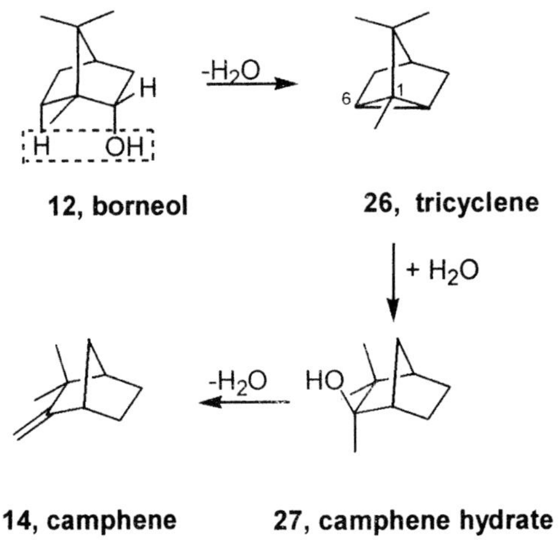

12, borneol
26, tricyclene
14, camphene
27, camphene hydrate

Scheme 11

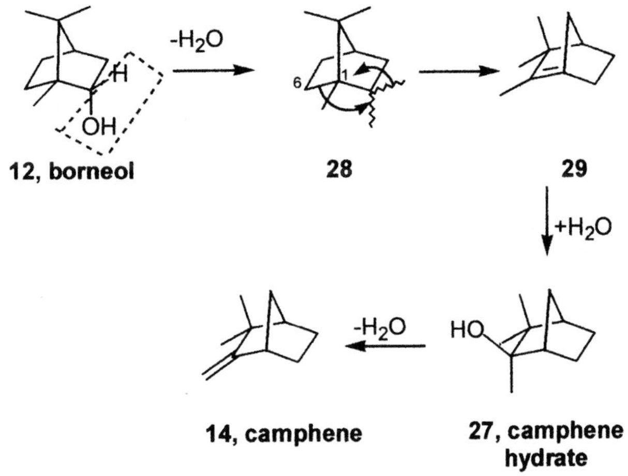

12, borneol
28
29
14, camphene
27, camphene hydrate

converting the bridgehead alkene **29**, which he knew from Bredt's work would be extremely unstable, to the final product camphene **14**. Meerwein was non-committal on the mechanism of the double bond shift needed to convert the bridgehead alkene **29** to camphene **14**.

I have shown it here in the spirit of that time as passing through camphene hydrate, but we would write the process today with a carbonium ion intermediate. Note that although Meerwein was aware of Bredt's prohibition against bridgehead double bonds, this did not deter him from invoking such a species as an intermediate. As we shall see, he never had to defend this (indefensible!) proposal.

Meerwein's rejection of the tricyclene hypothesis of Scheme 10 rested upon more than the usual arguments against stable intermediates then current in the pinacol wars. He rejected it because the camphene obtained from optically active borneol is itself optically active. *But tricyclene* (**26**) *is a bilaterally symmetrical molecule with a* *plane of symmetry and hence is achiral.* To preserve the tricyclene mechanism, one would have to violate a major tenet of symmetrization experiments, namely that a non-racemic chiral product cannot be formed from an achiral intermediate in a truly or effectively[37] achiral environment. In fact, for this mechanism to be correct would require a complete overthrow of the most basic laws of stereochemistry.

This very point had been made, almost in passing, in a paper by Semmler in 1902,[38] as Meerwein noted.[28] Discussing the mechanism of formation of camphene from borneol, Semmler said "... if however the formation of camphene goes through tricyclene, then the camphene must be inactive, which is in contradiction to experiment." What Meerwein did not note, however, was that Semmler soon backed away from his own seemingly categorical rejection of the tricyclene mechanism. Thus, in his monumental three-volume compendium of the chemistry of the essential oils published four years later,[39] Semmler discussed the mechanism only cursorily, using the tricyclene pathway without any reference to his previous exclusion of it! One can only conclude that for unstated reasons, Semmler had lost his conviction about the basic stereochemistry. Tiffeneau's paper proposing the α-elimination mechanism did not appear until three years later; Semmler, apparently unable to devise his own alternative, or too diffident to propose one, found no safe refuge and retreated to a well-populated (but vulnerable) fortress, the already discredited cyclopropane mechanism, of which the tricyclene hypothesis was, of course, an example.

Meerwein's 1914 paper[28] thus was his first attempt to avoid the (to him) unacceptable hypothesis of the tricyclene intermediate. In concluding that paper, he applied the α-elimination mechanism also to another terpene reaction (Scheme 12) with the confident assertion:

» *It appears almost superfluous to point out the conclusion that the considerations proposed in the forgoing work for the rearrangement of borneol* **12** *to camphene* **14** *[Scheme 11] can be transferred without modification to the processes [see Scheme 12] taking place in the formation of fenchene* **33** *from fenchyl alcohol* **30**. «

Scheme 12

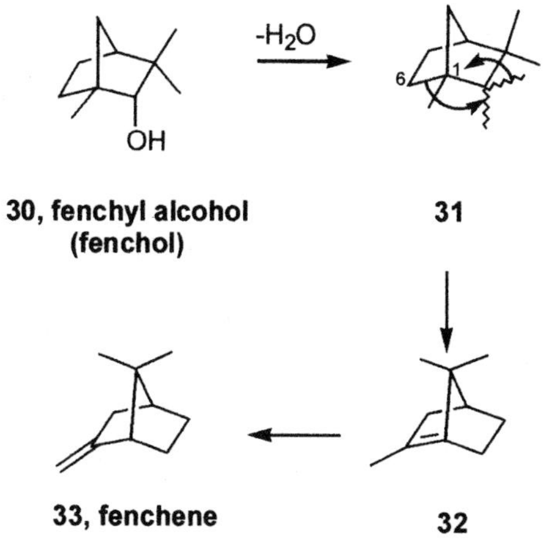

Meerwein's paper was submitted on April 4, 1914, four months before Germany, in fatal embrace with the other major nations of Europe, stumbled into the four-year dance of death called the Great War. Only in 1918, as the war ended, did Meerwein publish[40a] a further study of monocyclic model reactions, based upon a 1915 dissertation;[40b] he did not return to the problem of the mechanism of the terpene rearrangements themselves until 1920.

5.6 Ruzicka's Experiment and Support for the Tricyclene Hypothesis. Optical Activity Despite Achirality?

In the meantime, in 1918, Leopold Ruzicka, a 31-year-old professor (Nobel Prize in Chemistry, 1939) at the Eidgenössische Technische Hochschule in Zürich, in neutral, peaceful Switzerland, had become interested in these rearrangements and had published a paper[41] whose rationale is best rendered in his own words:

» *The question of which of the two explanations [cyclopropane intermediate vs α-elimination] would be found correct could now be simply decided in favor of the tricyclene, if one could demonstrate the Wagner rearrangement in a* tertiary *alcohol, because in such a case the intervention of a divalent carbon is not possible.* «

The difficulty of explaining such a result by the α-elimination mechanism, to which Ruzicka referred, is shown by comparing the Meerwein α-elimination mechanism for the borneol → camphene rearrangement (Scheme 11) with the corresponding α-elimination mechanism for the case of methylborneol **34** in Scheme 13. In the latter, one would have to eliminate methanol instead of water from a single carbon by breaking a C–C bond, a fundamentally different and unprecedented (hence according to Ruzicka "impossible") process. Moreover, one would have to *re-add* methanol

Scheme 13

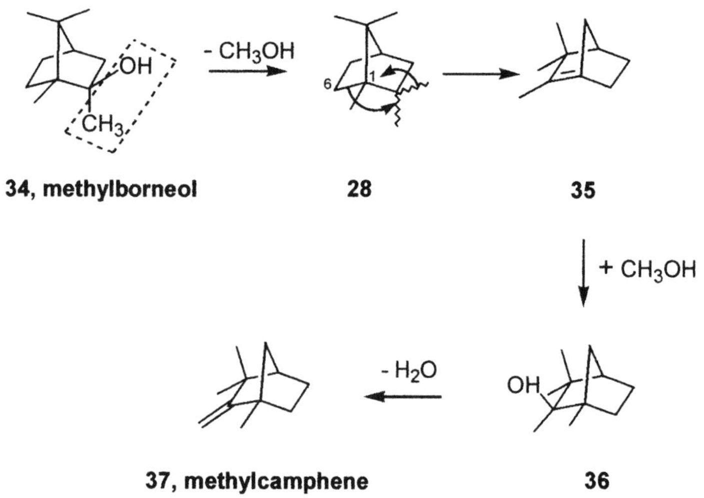

OH CH_3 $- CH_3OH$ 6 1 $+ CH_3OH$ $- H_2O$ OH

34, methylborneol **28** **35**

37, methylcamphene **36**

to the bridgehead olefin **35** by breaking a C–O bond rather than an O–H bond, which also would be unprecedented.

Ruzicka then carried out an ingenious experiment (Scheme 14). The readily available commercial forms of camphor **11** and fenchone **39** upon treatment with methyl Grignard reagent each gave a tertiary alcohol, methylborneol **34** from camphor **11**, and methyl fenchyl alcohol **40** from fenchone **39**. Each of these should give upon treatment with acid two major alkene products: one derived by simple dehydration to the exocyclic methylene derivative without rearrangement (**34** → **38**, **40** → **37**), and another derived by Wagner rearrangement (**34** → **37**, **40** → **38**). The choice of these two substrates has the beautiful consequence that the non-rearranged dehydration product of one is the rearranged product of the other! Thus, if Wagner rearrangement can occur, both reactions should give mixtures of the same two alkenes (although not necessarily in the same ratio). This vastly simplified the analytical problem, since both reaction mixtures upon ozonolysis would give mixtures of camphor and fenchone, which could be detected quantitatively by a procedure devised earlier by Wallach: camphor readily gives a semicarbazone under specified conditions, but fenchone, whose carbonyl group is sterically hindered by two adjacent quaternary carbons, does not. The fenchone is sufficiently volatile to be steam distilled and estimated gravimetrically, while the camphor semicarbazone is left behind.

Ruzicka showed that processing of each of the starting ketones (carefully shown to be free of cross-contamination at the outset!) in the sense of Scheme 14 led to mix-

Scheme 14

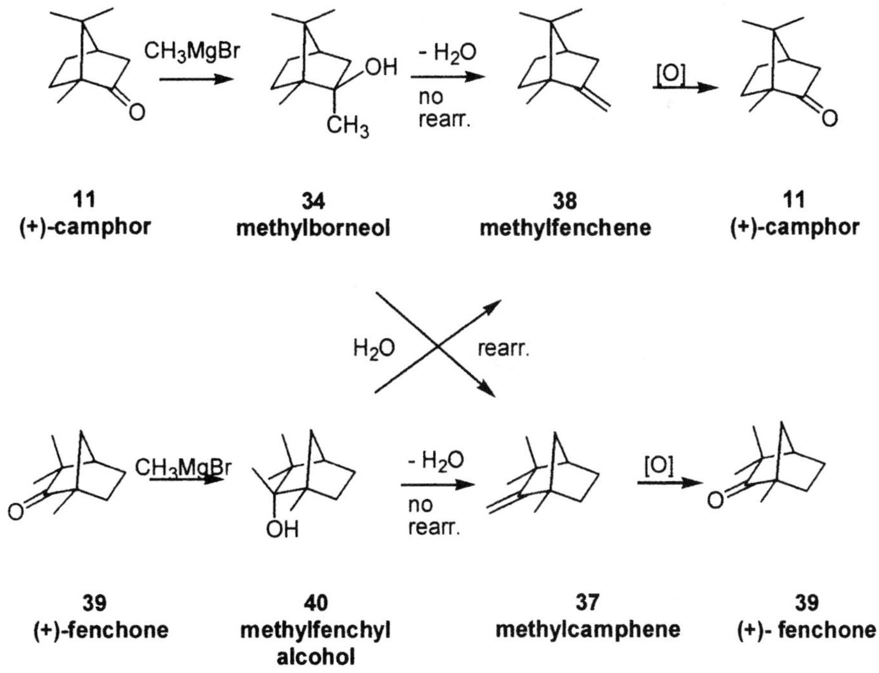

CH_3MgBr ; $- H_2O$, no rearr. ; [O] ; OH ; CH_3

11 (+)-camphor — **34** methylborneol — **38** methylfenchene — **11** (+)-camphor

H_2O ; rearr.

CH_3MgBr ; $- H_2O$, no rearr. ; [O] ; OH

39 (+)-fenchone — **40** methylfenchyl alcohol — **37** methylcamphene — **39** (+)- fenchone

tures of both methylfenchene **38** and methylcamphene **37**, which upon ozonolytic analysis, gave mixtures of camphor **11** and fenchone **39**. In other words, these results showed that Wagner rearrangement did occur readily in a bicyclic terpene in which the α-elimination mechanism must be considered highly improbable because the reactants were tertiary alcohols.

As a coup de grace to the a-elimination mechanism, Ruzicka argued that even the secondary cases to which Meerwein had applied the α-elimination mechanism were suspect (Scheme 11). Thus, Bredt and Holz[42] just the year before had shown that the thermal decomposition of 3-diazocamphor **41** (Scheme 15) gave a tricyclene derivative "camphenone" **43**, to which the earlier literature had incorrectly assigned a structure containing a carbon-carbon double bond. It was likely that the reaction leading to **43** passed through an intermediate divalent carbon species **42**. Ruzicka implied that this, rather than Wagner rearrangement, was behavior typical of such divalent carbon intermediates, but a moment's reflection shows the argument to be less than compelling, since a hypothetical Wagner rearrangement of **42** would face severe difficulties that were not present in the examples of Meerwein.[43]

Scheme 15

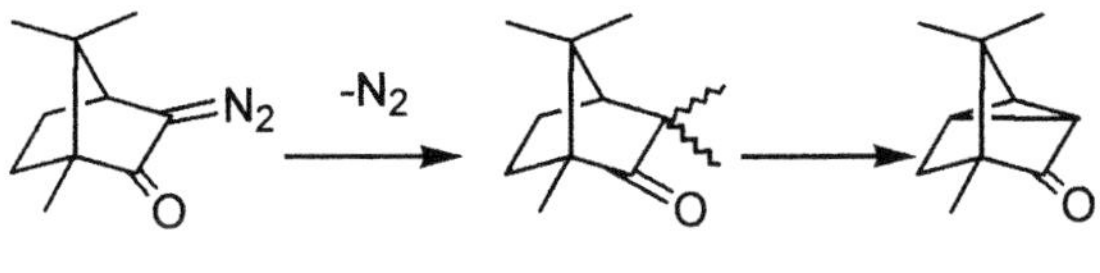

Ruzicka concluded with the words:

» ... *through my demonstration that the Wagner rearrangement also occurs with tertiary alcohols, the assumption of divalent carbon must be unconditionally rejected; thus, if one wishes to give a common explanation for the analogous transformation of the simple pinacol rearrangement and the Wagner rearrangement, a relationship that Wagner already had pointed out, the formation of a three-membered ring intermediate is the most simple explanation.* «

But what about the issue we already have mentioned, namely Semmler's (diffidently held) objection, repeated without equivocation in Meerwein's 1914 paper,[28] that the bilateral symmetry of the putative tricyclene intermediate in the cyclopropane mechanism of the borneol-camphene rearrangement would produce optically inactive camphene, in contradiction to experimental fact? In attempting to deal with this, Ruzicka fell short. It is difficult to paraphrase his tortuous and confusing argument, so I have simply repeated it here:

» *To be sure, Meerwein has collected several objections to the assumption of a three-membered ring in these reactions and considers the elimination of water from one carbon atom (following Tiffeneau, Scheme 11) to be the more likely explanation.*

But aside from the fact that the properties of methylborneol and methyl fenchyl alcohol compel the assumption of a tricyclene, and aside from the fact that the elimination of water from a single carbon atom is impossible in a tertiary alcohol, the Meerwein objections are untenable.

Meerwein based his first objection on a paper of Semmler[38] *in which it is stated that if the symmetrical tricyclene from borneol were the intermediate, an optically inactive camphene must result, but this is not the case. At that time (1902), Semmler assumed it to be self-evident that in the reduction of camphor to borneol an optically inactive hydroxyl carbon is formed. To the contrary it is to be noted that the carbonyl of camphor, lying between two optically active carbon atoms will be quite asymmetrically reduced, borneol would possess an additional optically active carbon atom, and therefore its tricyclene is to be considered an optically active compound.*

When Semmler made the objection above, asymmetric synthesis was not yet known. Besides, in his famous work[39] *on essential oils (of 1906), Semmler does not raise this objection; in contrast, he explains all these related reactions with the aid of a tricyclene.* «

Thus, Ruzicka was postulating that the presence of a so-called "optically active carbon" somehow could confer optical activity upon an achiral molecule. That he could propose this in 1918 illustrates just how insecure was the grasp of elementary stereochemical principles then, even by an outstanding investigator. Whether the paper was peer reviewed is not clear,[44] so that one cannot tell whether these misconceptions about stereochemistry were widespread at that time.

At this point we mention here for mnemonic purposes a further problem that Meerwein astutely recognized: although the camphene obtained in the reactions of bornyl and isobornyl derivatives described above was not usually completely devoid of optical activity, often it was partially racemized, the degree of racemization depending on the conditions of acid concentration, temperature, and time. Increases in any of these variables increased the degree of racemization. Thus, in alkaline medium, the hydrolysis of isobornyl chloride gave camphene of high enantiomeric purity, whereas the dehydration of isoborneol in strong acid a high temperature gave extensively racemized camphene.

Superficially, this might be thought to result from a combination of two mechanisms, one passing over optically active intermediates, and the other passing over the achiral tricyclene. However, as will become clear in the next section (5.7) this cannot be generally correct. The actual source of the partial racemization subsequently was to occupy Meerwein's close attention, but definitive solutions to the problem did not emerge for another 25–30 years. We shall discuss this further in section 5.9

5.7 Meerwein's Response

Ruzicka's report that Wagner rearrangements of tertiary alcohols occur, and his conclusion that they do not proceed by the α-elimination-divalent carbon mechanism, could hardly be contested. Meerwein now certainly must have realized that he had lost a battle. His favored α-elimination mechanism clearly was not general, and in fact, might be entirely false. Nevertheless, he was not convinced that the observations *required* a cyclopropane intermediate such as tricyclene, as Ruzicka had urged, and he remained adamant in his opposition to the tricyclene pathway. He had two bases for this: first, his perception that cyclopropane intermediates led to "difficulties" of interpretation of some of his monocyclic model rearrangements,[40a] (difficulties that we might not find compelling arguments today), and second, the insurmountable obstacle that the symmetrization required by the alleged tricyclene intermediate in the borneol → camphene rearrangement simply was not observed. For a while, the prospects for him to emerge victorious from this impasse may have seemed bleak. Yet we know, of course, that eventually he must have won out. After all, the class of rearrangements typified by the pinacol and bicyclic terpene transformations now bears the name "Wagner-Meerwein," not "Wagner-Ruzicka." How Meerwein reversed the tide of opinion running in favor of the tricyclene mechanism is instructive.

Essentially, he did it with two brilliant papers with Konrad van Emster.[45] In the first,[45a] he found additional evidence against the Tiffeneau-style divalent carbon species, and he devised a direct and irrefutable demonstration that tricyclene could not be the intermediate either. In the second,[45b] he discovered that the true intermediate is a "carbonium ion" (what is called today a "carbocation"). The latter result is the long-sought solution to the mysteries of the Wagner rearrangement, but its significance transcends that issue. A good case can be made that it was one of the most important discoveries of the 20th century in the field of organic chemistry. Below (Section 5.8), we discuss this second paper[45b] briefly, which has been analyzed in more detail elsewhere,[22] but first, we concentrate on the first van Emster paper[45a] in order to demonstrate the crucial role of symmetrization in Meerwein's thinking.

The first paper (1920)[45a] could hardly ignore the Ruzicka report,[41] but Meerwein's own new work had now convinced him to abandon some of his earlier positions. Recall that Ruzicka's two principal conclusions were that the Wagner rearrangements of the tertiary alcohols excluded α-elimination, and that *therefore*, the tricyclene mechanism was the most plausible alternative. Meerwein simply declined to offer an opinion on whether or not he agreed with the first of these. This was no doubt prudent, since he really had no basis for challenging it. Although as we have seen, Ruzicka's contribution was praiseworthy, Meerwein never gave it more than tacit acceptance. Instead, he put forward his own independent experimental refutation of the divalent carbon mechanism, and then he attacked Ruzicka's second proposal, namely the tricyclene mechanism.

Drawing upon the 1917 Bredt-Holz paper[42] (see Scheme 15) Meerwein and van Emster reasoned that it should be possible to generate **28**, the hypothetical divalent carbon precursor of camphene (see Scheme 11), by mercuric oxide oxidation of camphor hydrazone **44** (Scheme 16) via the mercury complex **45**. If **28** was indeed the key intermediate in the Wagner rearrangement, camphene should be formed as a major product. In fact, however, they found only traces of camphene. The major product (90% yield) was tricyclene (**26**),[46] apparently derived by insertion of the valences of **28** transannularly into a C_6-H bond. This confirmed (more convincingly) Ruzicka's point that such divalent carbon intermediates would not give Wagner products but instead would form tricyclenes.

Scheme 16

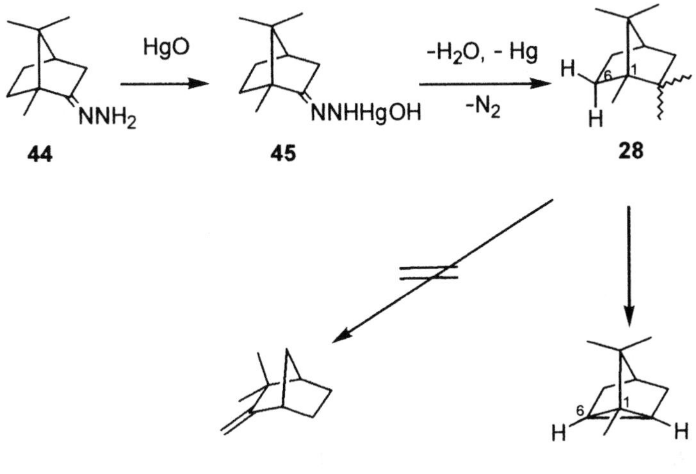

HgO
$-H_2O$, - Hg
$-N_2$
NNH_2
NNHHgOH
44 **45** **28**
14
camphene
26
tricyclene

In essential (but unadmitted) agreement with Ruzicka, Meerwein and van Emster[45a] now drew the reasonable conclusion that the Wagner rearrangement of, for example, borneol to camphene could not involve **28** as an intermediate (see Scheme 16), and they abandoned the Tiffeneau mechanism forthwith. They also recognized that serendipity had placed into their hands a simple method for the preparation of tricyclene, which although a known compound, was not found in significant quantities in the usual Wagner rearrangement mixtures and had not been readily available heretofore.

The most important of the experiments that now became feasible was a direct kinetic examination of the role of tricyclene in the Wagner rearrangements. Meerwein and van Emster showed that under the conditions in which isoborneol is completely converted to camphene (heating with 33% H_2SO_4 for several hours), *tricyclene does not react at all*. Morover, although isobornyl chloroacetate could be obtained from camphene by heating with chloroacetic acid (a reverse Wagner change), and it also could be obtained from tricyclene under the same conditions, the rate of the reaction with camphene was much greater than that of the reaction with tricyclene. These results,[45a] and similar findings by Lipp,[46b] were unequivocal refutations of the tricyclene hypothesis. As a final comment on Ruzicka's analysis of the mechanism, Meerwein dismissed the argument[41] that tricyclene from borneol should be optically active as "so unclear that it is superfluous to go into it any further."

Thus, despite having temporarily lost his way in advocacy of the incorrect Tiffeneau mechanism, and despite having to suffer Ruzicka's refutation of it, Meerwein finally was able to show that the tricyclene mechanism favored by Ruzicka and by

Semmler was incorrect. It is evident that Meerwein was certain that whatever might be wrong with the Tiffeneau mechanism, the tricyclene mechanism could not possibly be right. What was the source of his tenacity even in the face of adversity? I cannot help but believe that Meerwein found simply unacceptable the attempts by others to circumvent the principles of symmetrization experiments. In his mind, these were incontrovertible, and he clung to that conviction that until the problem was resolved.

5.8 Carbonium Ions (Carbocations) as Intermediates in the Wagner Rearrangements

These findings still left the problem of defining the nature of the intermediates in the Wagner rearrangements. A major step in the solution of this mystery came in the second Meerwein-van Emster paper.[45b] Detailed reviews of these developments have been given elsewhere,[22,47a,b] but it will be useful to summarize the results briefly here.

The crucial insight into the mechanism was the discovery[45b] that the rate of the rearrangement of camphene hydrochloride **46** to isobornyl chloride **49** was strongly dependent on the ionizing power of the solvent and was greatly enhanced by the presence of hydrogen chloride or Lewis acid catalysts. Meerwein and van Emster proposed that the key intermediates were the carbonium ions **47** and **48** (Scheme 17). Formally, the actual rearrangement step in the conversion of **47** to **48** may be thought of as a migration of C_6, accompanied by the C_6–C_1 bonding electron pair, from C_1 to C_2.

Scheme 17

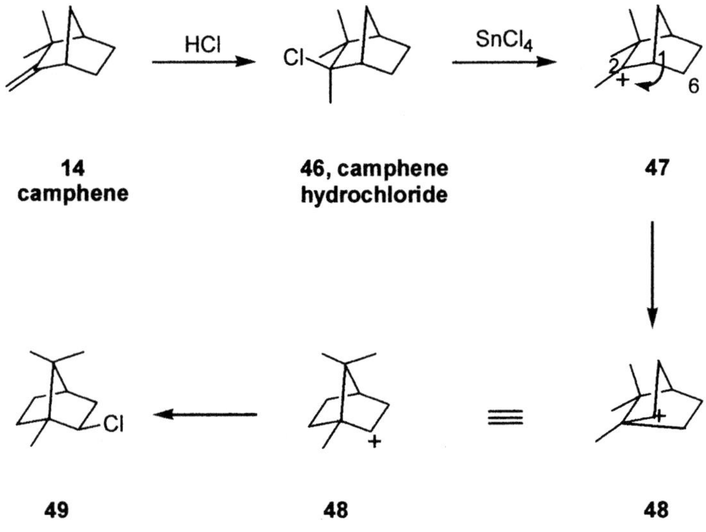

HCl
$SnCl_4$
14 camphene
46, camphene hydrochloride
47
49
48
48

One subtle question about the rearrangement has been the subject of intensive study and debate:[22,47c–l] Are there really two intermediates which exist as discrete "classical" entities **47** and **48**, or does the rearrangement pass through a single species, the "nonclassical" ion **52** (≡ **54**), as first suggested by Wilson and co-workers[47c] (Scheme 18). The problem is related, of course, to the issue of electrophile-alkene π-complexes we saw in Section 4.5.4, and to the even more general phenomenon of the neighboring group effect, which had been extensively elucidated by Winstein .[47i] Winstein and many others in the period 1946–1970 had deduced indirect but persuasive evidence that nonclassical ions could intervene in solvolytic reactions proceeding over carbonium ion intermediates in a wide variety of conventional nucleophilic solvents.

Scheme 18

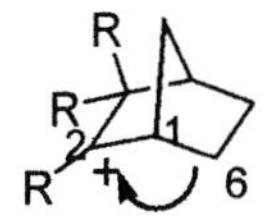

47: R = Me

50: R = H

48: R = Me

51: R = H

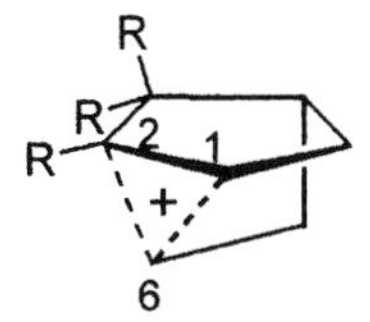

52: R = Me

53: R = H

54: R = Me

55: R = H

The discovery by Olah[47j] that in a series of non-nucleophilic, very strong acid solvents, such as SbF_5/SO_2 and $SbF_5/SO_2ClF/SO_2$, carbonium ions are no longer transients but persist for hours, made possible the direct spectroscopic observation and thermodynamic characterization of simple alkyl cations as stable species. By this means, Olah and numerous other workers deduced the structures of a wide range of carbonium ions. One of the most significant forward steps was the development by Saunders of the method of isotopic perturbation of equilibrium, observable in the ^{13}C NMR spectrum. The application of this ingenious technique provided compelling evidence that the parent 2-norbornyl cation, under such acidic conditions, indeed has the bridged structure **53** (≡ **55**) and is not a rapidly equilibrating mixture of **50** and **51** (Scheme 18).[47k,l]

5.9 Racemization of Camphene and Related Symmetrization Phenomena.[22]

As we have seen, partially or fully racemized isobornyl derivatives are obtained from the rearrangement of enantiomerically pure camphene. Meerwein's attempts to understand this phenomenon passed through some just plain erroneous structural assignments, some bizarre speculation, and some unclear but extremely original thinking. At least part of his difficulty, I think, was that he was decades ahead of most of his contemporaries in his ability to *discern a problem*, but the experimental and theoretical tools needed to solve it completely did not yet exist. Sometimes, one can hope to invent, on the spot and ad hoc, the necessary new tools, but often, the investigator can only go so far as to bring to the world's attention the fact that a new and unexplained phenomenon remains to be elucidated. Further progress then may have to await the arrival of the needed methods in slow advances across a broad front of the discipline. In the meantime, the pioneer may find only frustration in attempts to find the answer using the current inadequate means.

Meerwein's attempt to explain the racemization of isobornyl chloride, which is discussed below, received what he interpreted to be an encouraging stimulus from a study with Wortmann[48] of another system, the rearrangement of "α-camphordichloride" **56**, a product of the action of PCl_5 on camphor (Scheme 19). Treatment of this dichloride with stannic chloride in benzene led to an isomer, called "β-camphordichloride," to which Meerwein and Wortmann assigned the structure **59**, largely on the basis of the observation that reduction with sodium and alcohol gave camphane **60** in nearly quantitative yield.

Scheme 19

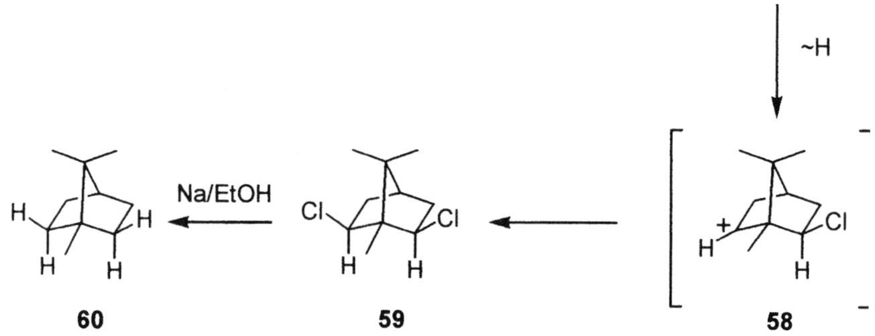

These authors visualized the rearrangement as taking place in the intermediate carbonium ion **57** by a new kind of reaction, a *non-vicinal* migration of a hydrogen from

C_6 to C_2.[48b] We return in a moment to Meerwein's formulation of the structure of "β-camphordichloride,", which in fact is incorrect, but it is necessary to keep it in mind as having strongly influenced Meerwein's thinking about the self- racemizations of camphene and isobornyl chloride, and the racemization in the rearrangement of camphene hydrochloride to isobornyl chloride.

Meerwein and van Emster[45b] already had observed the racemization of isobornyl chloride **49**, a process that Meerwein and Montfort[49] now studied in some detail. It became quite clear that this reaction, like the rearrangement of camphene hydrochloride to isobornyl chloride (**46** → **49**, Scheme 17, Section 5.8), involved carbonium ion intermediates: the rate was increased by polar solvents and by catalysts such as cresol, $SnCl_4$, or $SbCl_5$.

Meerwein and Montfort now proposed two alternative mechanisms for the racemization of isobornyl chloride **49**. One considered the possibility that the racemization took place without rearrangement of the the molecular structure, so that the chlorine atom remained attached to the carbon to which it was originally bound. In order for racemization to occur, it then would be necessary for inversion to take place at C_1, C_4, and C_2, that is, *all three* of the stereocenters of isobornyl chloride. Overall, the process would involve the equivalent of bending the bridge carbon (C_7) *through* the molecule in order to invert C_1 and C_4. Meerwein did not give any indication of how likely he thought this possibility was, but in retrospect, one perhaps could say that so little was known in 1924 about the properties of carbonium ions that the reaction could not be dismissed out of hand. Of course, today we would consider it energetically improbable.

The second hypothesis for the racemization of isobornyl chloride **49** grew directly out of the Meerwein-Wortmann work. Having put forward the ionization-6,2-shift mechanism for the "α-camphordichloride"-to-"β-camphordichloride" rearrangement (allegedly **56** → **59**) shown in Scheme 19, Meerwein and Montfort proposed the analogous 6,2-shift shown in Scheme 20. The key step is the 6,2-hydrogen shift which interconverts the ion-pairs **48a** and **48b** and permits the interconversion of one enantiomer of isobornyl chloride **49** to the other **49′**.

This formulation showed a surprising degree of sophistication for its time in that it explicitly took into account the fact that in these solvents, which ranged in polarity from petroleum ether to nitromethane, counterions could be associated with the carbonium ion. (Meerwein showed Cl^-, but we today would think it more likely that they would be Lewis acid-base complex ions such as $SnCl_5^-$).

Before we discuss what was really happening in these rearrangements, we should pause to notice one additional striking insight of Meerwein: in his own words, "the autoracemization in the latter case does not, as has always been assumed, *pass over a symmetrical intermediate*." [emphasis in the original]. Of course, his claim that the autoracemization "*does not*" pass over a symmetrical intermediate is a bit too strong. In order for this to be the case, the counterion must avoid taking up a position midway between C_2 and C_6 of cation **48** simultaneously with the arrival of the C_6 hydrogen at the midpoint of its passage to C_2 (Scheme 20), or it must avoid receding to "infinite" distance before the hydrogen shift occurs). Either of these contingencies would correspond to a symmetrical intermediate. Therefore, a more accurate way of

Scheme 20

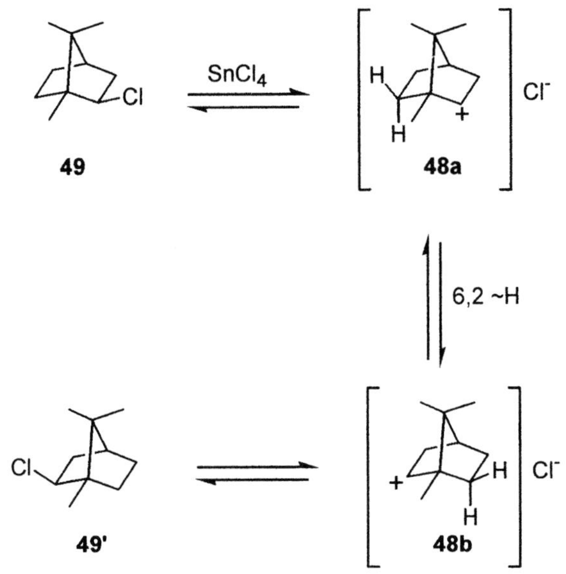

SnCl$_4$ 49 48a Cl$^-$ 6,2 ~H 49' 48b Cl$^-$

stating the matter would be that the automerization *can* occur without the intervention of a symmetrical intermediate. To appreciate the remarkable prescience of Meerwein's thinking, we must realize that it would be another thirty years before anyone was to point out explicitly that reaction symmetrization does not require a molecular symmetrical point on the pathway and to provide a case in which, in fact, racemization *must* occur without such a species. It would take an additional decade beyond that before the recognition of the full implications of this generalization for the applicability of the principle of microscopic reversibility to reaction mechanisms (see Section **5.11**).

The Meerwein-Montfort proposal in Scheme 20 of 6,2-hydrogen shift for the racemization of isobornyl chloride received a serious set-back, when Houben and Pfankuch[50a] showed that "β-camphordichloride," the rearrangement product of "α-camphordichloride" did not have the structure 2,6-dichlorocamphane **59** shown in Scheme 19, but instead was actually 2,4-chlorocamphane **65**, which they proposed to be formed by a Nametkin rearrangement (vicinal methyl shift), as in Scheme 21. (For didactic reasons, I have shown the rearrangement as occurring by Meerwein-style carbonium ion mechanisms, although neither Nametkin himself[50b] nor Houben and Pfankuch used that notation). Houben and Pfankuch had proposed earlier[50c] that Nametkin shift provided a preferable alternative to 6,2-shift as the long-sought explanation of the racemization phenomena in the camphene series: "*This is the elucidation of the observed racemization*, not the Meerwein theory of a 2,6-oscillation of a halogen atom" [emphasis in the original]. In the same year 1931, Bredt[50d] also recognized that Nametkin rearrangement would lead to racemization in that system, and claimed that, by the right of priority, he should have to himself the further investigation of the camphene racemization problem. This did not win the acceptance of others, particularly Houben,[50c] who claimed to have recognized the same explanation earlier.[50e]

Scheme 21

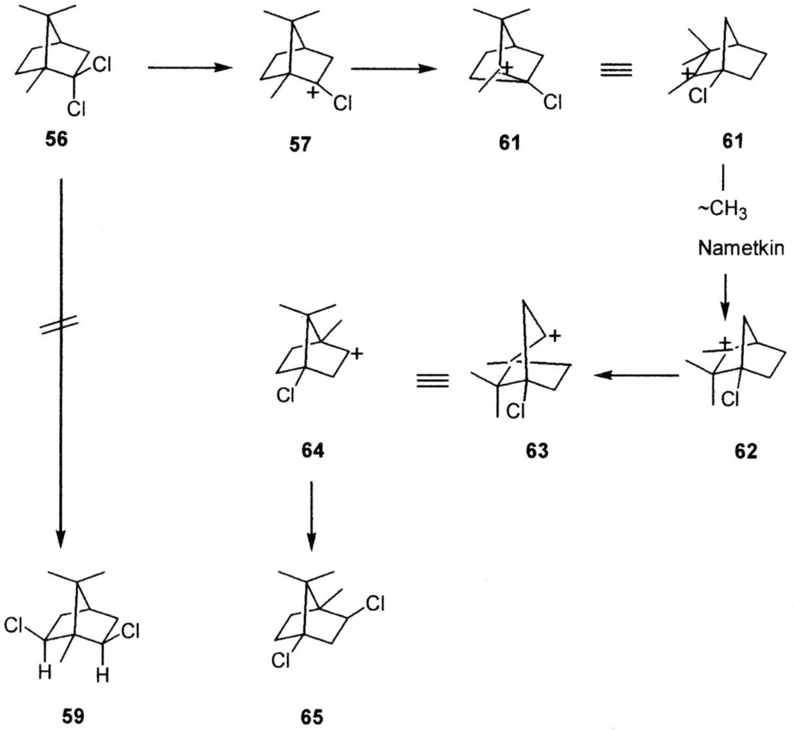

At this point, one cannot help but pause to admire the level of ingenuity, insight, and sophistication of the participants in the debate over these intricate rearrangements. Although civility barely survived some of the exchanges of views, the passionate attachments of the protagonists to their specific proposals made clear the agreement of all the parties that their subject was both fascinating and significant. In my opinion, this problem of the racemization in the camphene series marks an intellectual pinnacle of mechanistic research in the first third of this century, and the list of contributors to its resolution – Meerwein, Nametkin, Bredt, Houben – includes some of the most perceptive organic chemists of that era.

It is ironic that, although the experimental basis for Meerwein's proposal of the 6,2-hydrogen shift now had been refuted, the conclusion by Houben and Pfankuch[50a] that the Nametkin rearrangement was *solely* responsible for the racemization phenomena in the camphenehydro-isobornyl system was shown decades later to be incorrect also. The application of ^{14}C tracer studies to these rearrangements showed that the Nametkin rearrangement and the 6,2-shift mechanisms *both* were operative.[50f,g] Moreover, the ^{14}C tracer method showed the occurrence of both 3,2-

(vicinal) *hydrogen* shifts and 6,2-hydrogen shifts in the unsubstituted norbornyl cation formed under solvolytic conditions from norbornyl derivatives,[50h] and deuterium tracer studies[50i] showed 6,2-hydrogen shifts in the cations formed in the acid-catalyzed dehydration of fenchol.

Among the lessons for us in the history of the camphene racemization, we might include one summarized in the sour comment that Meerwein ultimately turned out to be right for the wrong reasons. This formulation, however, trivializes the matter. It seems to me much more significant that his *imagination* could produce ideas that eventually proved to be fruitful, *even though the original stimulus came from incorrect observations*. There is no doubt in my mind that Meerwein's idea of 6,2-hydrogen shift, a non-vicinal rearrangement, was considered misguided, not to say a bit wild, by many at the time. Well, it *was* a bit wild, but we must keep in mind that on the frontier of a truly new and unexplored territory, we really don't know what to expect. In those circumstances, it is wise to be ready for the unusual.

A second lesson is that, in retrospect, all of the participants in this story failed to anticipate the extraordinary variety of rearrangements that carbonium ions can undergo. We saw this also in the dienone-phenol story in Chapter 4, where attempts to find a governing mechanistic pattern failed because Nature had ingeniously designed a set of closely competitive carbonium ion reaction pathways.

Finally, a third lesson is that it is often difficult to apply a true *coup de grace* to a mechanistic proposal. We take comfort in the oft-repeated maxim that we can never prove a mechanism right; we can only prove it wrong. This is true in the abstract sense, and following this protocol, we may think that, by some clever experiment, such as the Houben study of Scheme 21, we have definitively excluded one mechanistic possibility. But as a practical matter, even the proof that it is wrong may be elusive.

5.10 The Favorskii Rearrangement. Symmetrization yes, but via Which Intermediate?

Symmetrization experiments often are designed as tests for a particular candidate symmetrical intermediate. A positive outcome, namely symmetrized product formed under conditions where the unsymmetrized product is stable, can give the investigator great joy and conviction that the candidate intermediate is actually responsible for the outcome. This reaction, of course, must be resisted, for two reasons: First, it is necessary to show that symmetrization of the reactant has not occurred prior to reaction. Second, even if that requirement is satisfied, there remains the possibility that the candidate intermediate is not the only symmetrical species that can intervene on the reaction pathway.

Both of these points can be illustrated by many examples, from which I have chosen some influential studies on the Favorskii rearrangement,[51–55] beginning with the well known isotopic position-labeling experiment of Loftfield (Scheme 22).[51]

Scheme 22

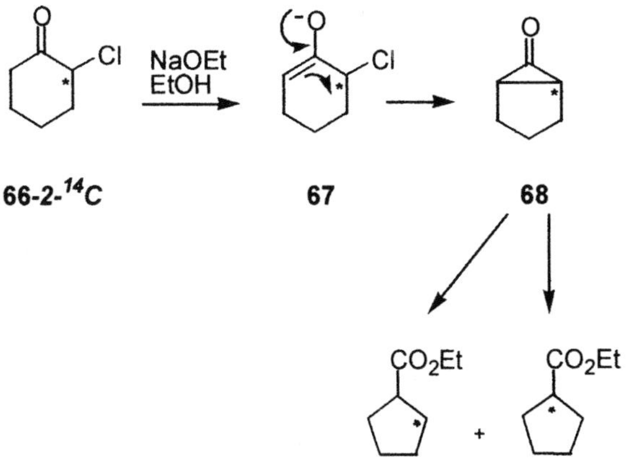

66-2-^{14}C 67 68

69-2-^{14}C 69-1-^{14}C

Loftfield showed that the rearrangement of the specifically labeled 2-chlorocyclohexanone **66**-2-^{14}C gave ethyl cyclopentanecarboxylate **69** in which the isotopic label was equally distributed between positions 1 and 2. In the original communication,[51a] he favored the symmetrical cyclopropanone intermediate **68**, formed by ring-closure of the enolate **67**, as the most likely explanation of this result. He was aware, however, that the observed symmetrization in the product could conceivably have resulted from an allylic rearrangement in the enolate or in the enol **70**-2-^{14}C of the starting ketone (Scheme 23). This potential side-reaction would be independent of the Favorskii rearrangement itself, and in that case, the experiment would carry no information about whether the rearrangement involves a symmetrical intermediate. In a subsequent full paper,[51b] he reported the completion of the demanding control experiment necessary to distinguish between these alternatives: recovery of unreacted starting material from a partially reacted run and demonstration that the 2-chlorocyclohexanone still contained the label exclusively at C_2 and not at C_6. The symmetrization observed in the rearrangement product thus could not be attributed to prior symmetrization of the reactant.

Scheme 23

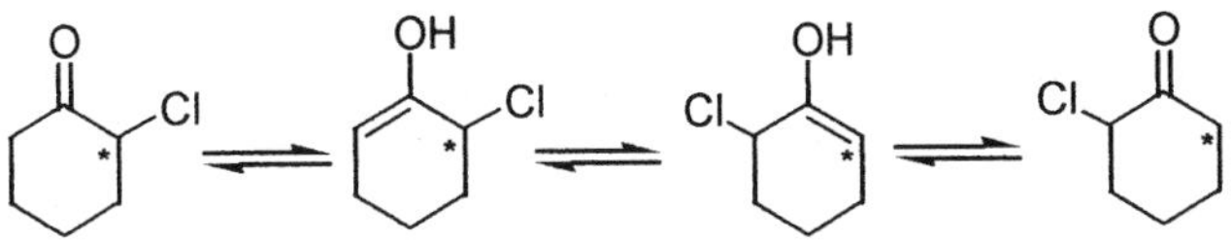

66-2-^{14}C 70-2-^{14}C 70-6-^{14}C 66-6-^{14}C

The second concern was whether the cyclopropanone **68** is the actual intermediate responsible for the symmetrization. Loftfield entertained the possibility that the cyclopropanone would be better represented as an oxyallyl, but he apparently thought such a species **71** would be simply a resonance form of the cyclopropanone **68** (Scheme 24). However, we would say today that since interconversion of **68** and **71** would have to be accompanied by drastic changes in geometry, this proposal is untenable, and **71** would have to be considered a true alternative symmetrical intermediate. In fact, just such an oxyallyl intermediate was proposed at that time by Aston and Newkirk[54] and later supported on theoretical grounds by Burr and Dewar.[55]

Scheme 24

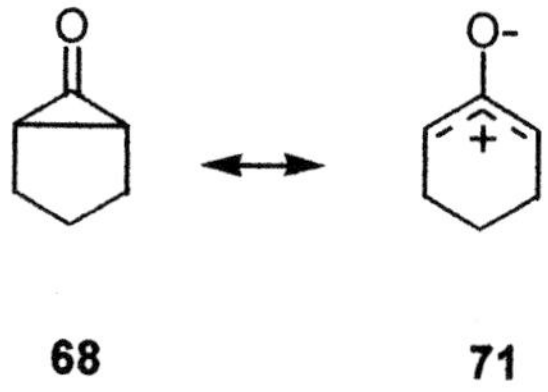

O O-

+

68 **71**

Experimental evidence that at least two different mechanisms exist for the Favorskii rearrangement first became clear from work by Stork and Borowitz[52] and by House and Gilmore.[53] Stork and Borowitz argued that if, as is implied in Scheme 22, the ring-closure of the enolate **67** to the cyclopropanone **68** is an intramolecular S_N2 reaction, it should occur with inversion of configuration at the original chlorine-bearing carbon. Ring-opening ethanolysis of either of the hypothetical cyclopropanones does not jeopardize the configuration established in the ring-closure step, and thus, this mechanism applied to the stereoisomeric chloroketones **72** and **75** should result in complete stereospecificity in the sense shown in Scheme 25.

Scheme 25

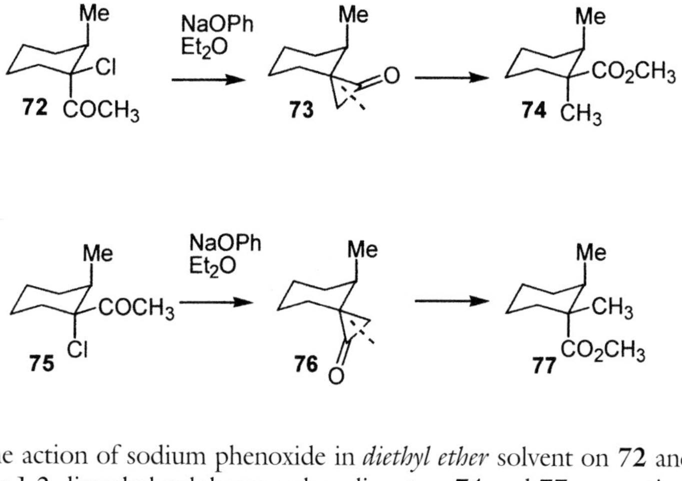

Me Cl **72** $COCH_3$ NaOPh Et_2O Me O **73** Me CO_2CH_3 **74** CH_3

Me $COCH_3$ **75** Cl NaOPh Et_2O Me **76** O Me CH_3 **77** CO_2CH_3

Experimentally, the action of sodium phenoxide in *diethyl ether* solvent on **72** and **75** gave the isomeric 1,2-dimethylcyclohexancarboxylic esters **74** and **77**, respectively (Scheme 25), with the predicted stereospecificity to a high degree, which strongly supported the cyclopropanone mechanism.

However, House and Gilmore[53] carried out a study of the solvent-dependence of the Favorskii rearrangement of the same pair of epimeric chloroketones. With sodium methoxide in solvent dimethoxyethane, the rearrangement was again stereospecific (**72** → 95 % of ester **74**), but in methanol, **72** gave a mixture of both stereoisomeric esters, 51 % of **74** and 41 % of 77. They interpreted these results and other similar findings as indicative of the intervention of an oxyallyl intermediate, e.g., **78** and/or **79** (Scheme 26), in which the geometry at the reacting stereocenter had become planar. It would appear then that both mechanisms can occur under the proper conditions.

Scheme 26

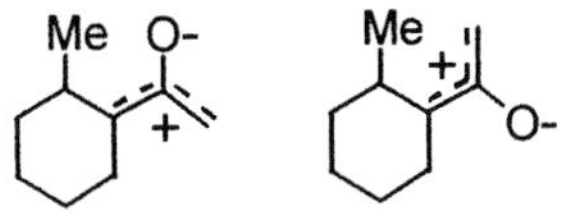

Me O-
+
Me
+
O-

78 **79**

This raises a further subtle question of the detailed mechanism of the randomization in the non-stereospecific version of the Favorskii rearrangement. Two possibilities present themselves: Does the chloroketone give the cyclopropanone first, which is stereorandomized by a fast reversible ring-opening to the oxyallyl, in competition with a slower alcoholysis to ester product, or does the chloroketone give the oxyallyl directly as the initial intermediate, which then ring-closes to the cyclopropanone? This point has been widely debated,[56] and a thorough discussion here would take us far afield, but it now appears that in at least some cases, the second of these alternatives, direct formation of the oxyallyl, predominates.[57]

5.11 Symmetrization yes, but Is there a Symmetrical Intermediate? Racemization Machines with no Achiral Parts. Reaction Symmetrization Without Molecular Symmetry

In transition state theory,[58] which is the framework of the forgoing discussion, if a symmetrical intermediate is on the reaction pathway, symmetrization of products must result. Is the converse also true, that is, if symmetrization is observed, is a symmetrical intermediate required? As we have seen (Section 5.9), Meerwein recognized that there could be reaction symmetrization without a symmetrical intermediate, but thirty years later, Mislow gave a more explicitly articulated formulation of this idea for a case in which no conceivable symmetrical pathway was accessible.[59a,b] This insight followed on an analysis[59c] of the phenomenon of meso character in conformationally mobile molecules.[59d]

A classical example of dynamic meso character is the interconversion of the pairs of enantiomerically related conformations (e.g., A and C) of *meso*-tartaric acid (Scheme

27). The reaction may be plausibly imagined to take place by way of the symmetrical achiral intermediates B and D, which, respectively, have a center and a plane of symmetry.

Scheme 27 (a = CO_2H, b = OH, c = H)

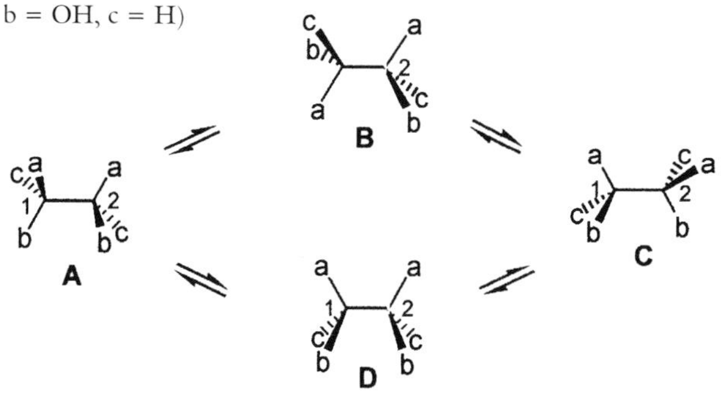

Is it possible to imagine a molecule in which conformational interconversions of enantiomers, and hence racemization, occurs when the formation of an achiral intermediate is *prohibited*? Mislow pointed out that the hypothetical molecule (of which a derivative was subsequently synthesized) shown in Scheme 28 corresponds to this situation.

Scheme 28

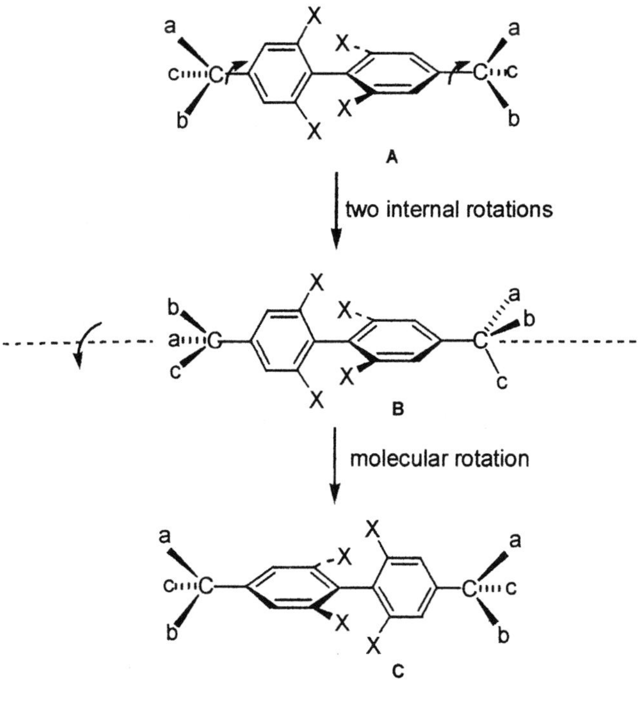

two internal rotations

molecular rotation

The conversion of **A** to its enantiomer **B** could be effected by, for example, internal rotations of the para substituents in the same direction. In such rotations, no achiral conformation would be encountered. Although an achiral conformation could be achieved if the two aryl rings were to come into planarity, *this could not occur if the substituents X were large enough*. Mislow recognized that the existence of this molecule (even if only in thought at the time) established the general principle that racemization can occur even when an achiral intermediate is inaccessible.

This has important implications for the applicability of the principle of microscopic reversibility (pmr) in reaction mechanisms, as was first pointed out more than a decade later by Burwell and Pearson (who apparently were unaware of the discussions by Meerwein and by Mislow),[60] and subsequently by Salem and co-workers.[61] Both of these papers[60,61] demonstrated the consequences of their ideas for the properties of energy surfaces of symmetrization reactions, including not only enantiomerizations but also isotopic exchanges. Their major conclusions may be summarized as follows: reaction symmetrization requires either (1) a one-path mechanism through a symmetrical species on an energy surface with forward and reverse trajectories related by reflection symmetry, or (2) a two- (or more) path mechanism on an energy surface with corresponding points on one or more *pairs* of competing paths in which the members of the pairs are related both structurally and energetically by C_2 symmetry operations. Overall microscopic reversibility is maintained in both cases.

Following Burwell and Pearson,[60] we show both symmetrical and unsymmetrical cases in Figure 2. Symmetrical one-path mechanisms for symmetrization reactions are represented in Figure 2, a and b. Whether the reaction trajectory passes over one transition state (Case a), or over two with a metastable intermediate (Case b), the reverse pathway is the mirror image of the forward one, and the structure of the reaction complex at the mid-point of the reaction coordinate has reflection symmetry. A simple example of Case a would be the S_N2 radiobromide exchange reaction between CH_3Br + $^*Br^-$ with a single symmetrical transition state; Case b would represent the profile of, for example, the exchange of oxygen in the reaction of CH_3CHO with $H_2{}^{18}O$ through the symmetrical metastable intermediate hydrate $CH_3CH(OH)(^{18}OH)$.

Unsymmetrical pathways are shown in Fig. 2c and d. Although the pmr requires a symmetrical mid-point in a symmetrization reaction occurring by a one-path mechanism, this restriction is lifted, under certain conditions, in a reaction that can occur by two parallel pathways. These conditions are that the second path must be constructed by *reflecting the first (unsymmetrical) path (e.g.*, the full heavy curve in Fig. 2c) *in a plane at the mid-point*. It is seen that this produces the dashed curve of Fig. 2c and that the *sum* of these two curves (not shown) is a symmetrical one. The two curves can be taken to represent the projections of two multi-dimensional pathways in coordinate space which are mutually interconverted by a C_2 symmetry operation about an axis through the mid-point. This relationship guarantees that not only the transition states but each pair of points so interconverted must have the same energy. The same protocol applies in a reaction with one or more intermediates as is outlined below(Case d).

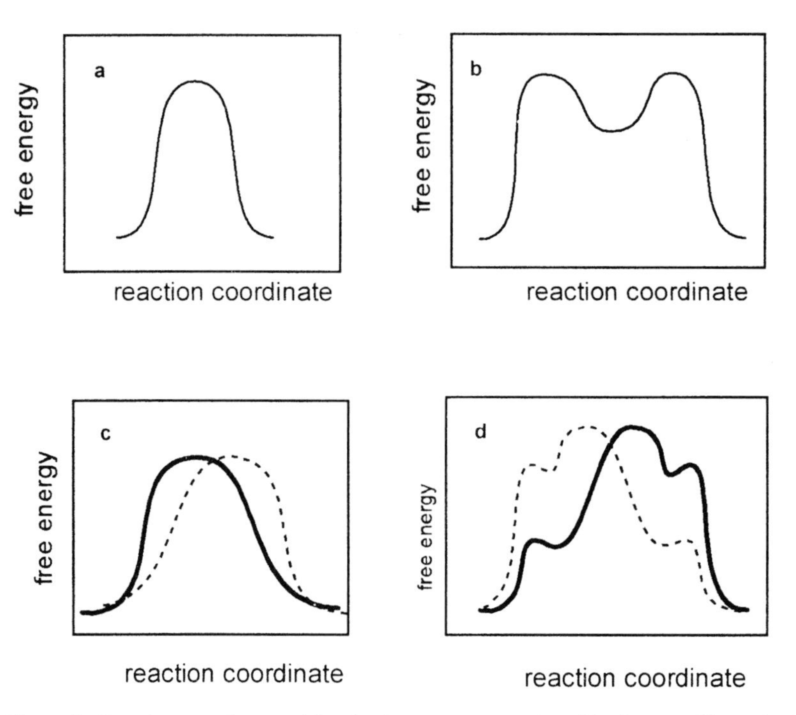

Figure 2. Reaction coordinates with reflection symmetry at the midpoint (a and b) and without reflection symmetry (c and d).

Case c

A simple example of case c is given by the isotopic exchange reaction of Scheme 29. In Scheme 29, the 180° relationship of the attacking nucleophile and the leaving group conventionally required by the S_N2 mechanism is imposed. The trajectory of the incoming bromide in this drawing is imagined to lie between the substituents H and Me, but it will be seen that regardless of what trajectory is chosen, the presence of the stereocenter adjacent to the reaction site prevents a symmetrical transition state. Moreover, the reverse pathway **C** → **B** → **A** is not microscopically the reverse of the forward one **A** → **B** → **C**, since the approach of the incoming bromide in the reverse pathway now is not between the substituents H and Me but rather alongside the Et group. Nevertheless, this situation does not violate pmr, because by the analysis in the preceding paragraph, there must be an opposing reaction whose trajectory is related in coordinate space to the forward one by a C_2 rotation and which has a transition state of equal energy. A moment's reflection reveals such a path, which is **A** → **C′** → **B′** → **A′** (Scheme 29). The primed species are the same as the unprimed ones except for the position of the isotope. Scheme 29 shows the C_2 relationship of the two pathways. Thus, the rate constants in the two directions are equal, as is required for an isotopic exchange reaction (disregarding any zero-point energy difference of the isotopomers) or for an enantiomerization. Although the point is made in Scheme 29 with the example of two opposing pathways, the same analysis holds for all the many trajectories imaginable by rotation about the C_1–C_2 bond of the reactant: each forward reaction will be matched by a reverse reaction of equal rate.

Scheme 29

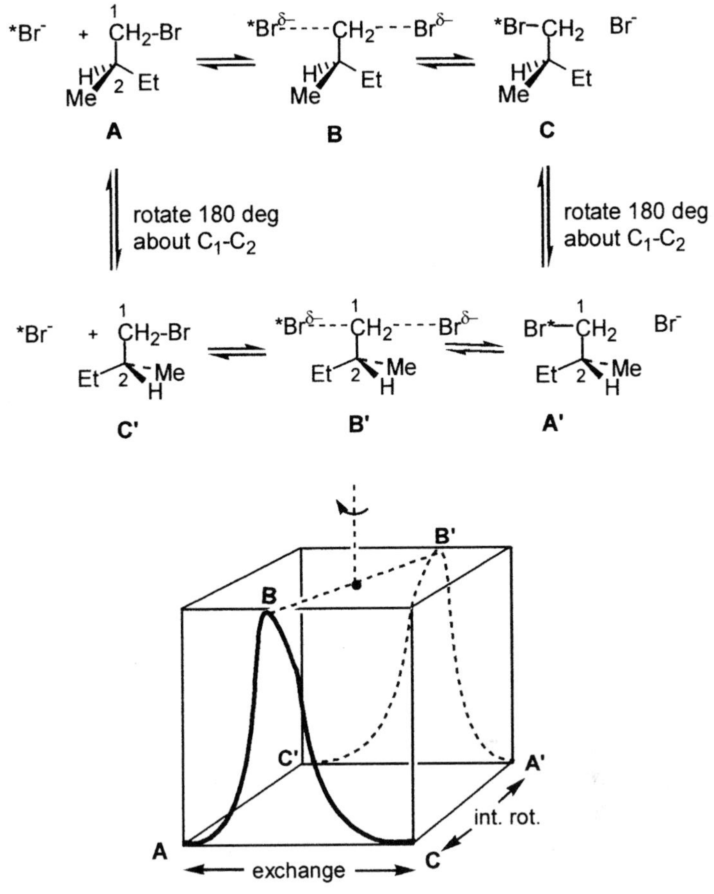

5.11.1 Case d. Symmetrization via a Multi-step Mechanism Without Symmetrical Intermediates

With the reader's indulgence of the author's self-referential prerogative, we find an interesting application of the multi-step type shown in case d of Fig. 2 in the observed interconversions of Scheme 30.[62a–c] The reactions of Scheme 30 are thermal Cope rearrangements which occur at or near a temperature of 471 K and interconvert acetylenes **80** with allenes **81**. Heating any of the four isomers produces a mixture containing the other three in addition to remaining starting material. The major mechanistic point at issue is: are the rearrangements concerted, or do they occur via cyclized biradical intermediates of the general type **82** (Scheme 31).

Scheme 30

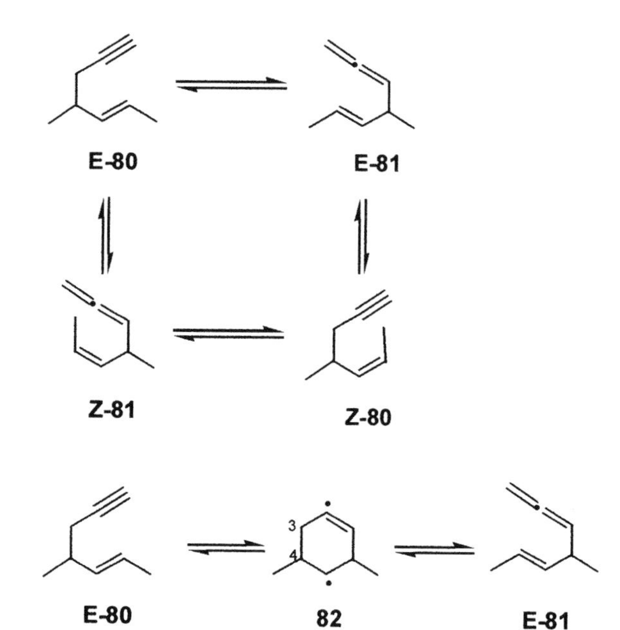

Scheme 31

One experimental method for detecting such biradical intermediates uses enantiomerically enriched reactant. The details[62a–c] are quite complicated, but for the present purpose it suffices to be aware that racemization of reactant or products in such a system could not occur as long as the rearrangements are all suprafacial or all antarafacial. Because the suprafacial and antarafacial pathways would lead to enantiomeric products, partial racemization would result if the reaction occurred with mixed faciality. In that case, since the rearrangements are all reversible, the enantiomeric purity of all the components eventually would decline to effectively zero after a sufficient number of reaction half-lives.

In a competition between a concerted [3,3]-sigmatropic suprafacial rearrangement, which is orbital symmetry-allowed and sterically favorable, and a concerted antarafacial one, which is not only orbital symmetry-forbidden but also sterically difficult, one would expect the suprafacial process to dominate. Thus, a concerted reaction should give complete preservation of enantiomeric purity. However, if a biradical intermediate such as **82** could intervene, it would provide a sterically plausible mechanism for formation of products of mixed faciality via conformational isomerization (Scheme 32), which would lead to racemization. It should be clear that the conformational isomerizations of the biradicals are not themselves racemizing events. The biradicals are individually *chiral*, regardless of their conformation or of the relative configuration of the two methyl substituents, and their chirality survives any conformational interconversions intact. Rather, it is because they can act as crossover points between adjacent suprafacial and antarafacial rearrangement pathways that their conformational interconversions ultimately can produce racemization of the reactant or product.

Scheme 32

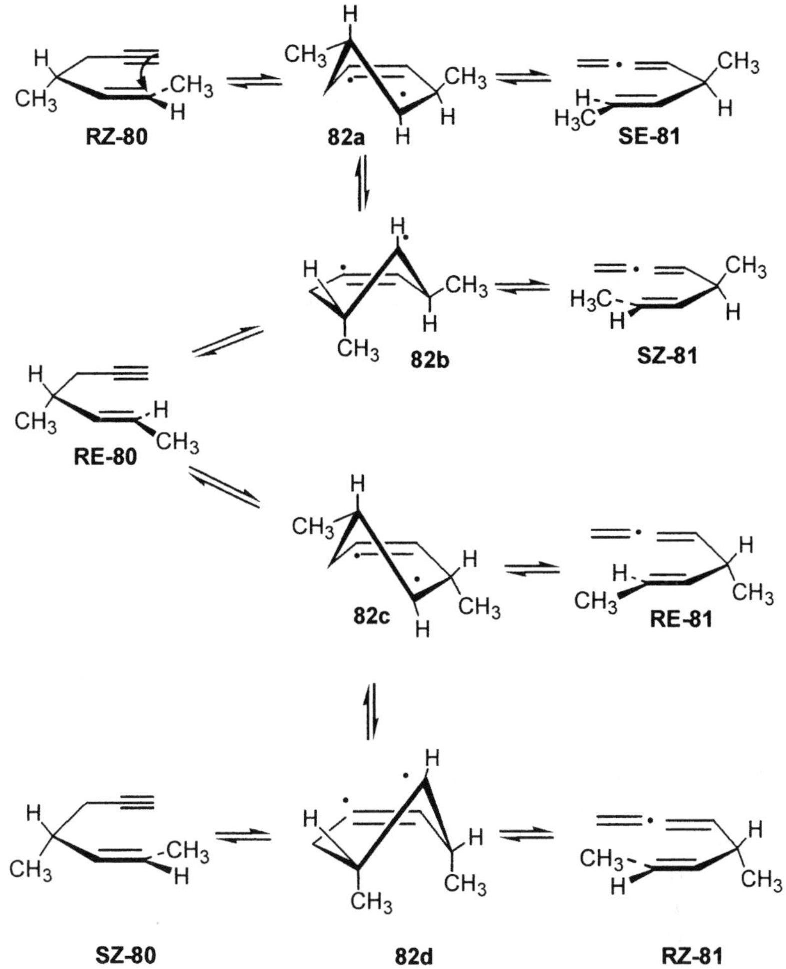

RZ-80 82a SE-81

82b SZ-81

RE-80

82c RE-81

SZ-80 82d RZ-81

These events can be seen in Scheme 32, a *partial* display of all the rearrangements, which nevertheless shows enough to illustrate the crossover mechanism. Thus, starting with the acetylene RZ-**80** (upper left), a direct concerted suprafacial rearrangement would produce the allene SE-**81**. Alternatively, the latter could be formed in a stepwise rearrangement via biradical **82a**, provided that **82a** suffers bond cleavage faster than it undergoes conformational isomerization to biradical **82b**. Inspection of the scheme now shows that once the biradical conformational isomerization channel is open, a path is made available for ultimate formation of the enantiomeric allene RE-**81**. The latter product is the same as would have been obtained from a hypothetical antarafacial rearrangement of the starting RZ-**80**. The biradical pathway also opens pathways for the enantiomerization of the starting acetylene, RZ-**80** to SZ-**80**. Note that Scheme 32 also provides a mechanism via conformational isomerization for the stereochemical equivalent of torsion about the double bond of the acetylene, for example RZ-**80** → RE-**80**, or the non-allenic double bond of the allene, for example, SE-**81** → SZ-**81**.

Instead of writing down all of the stereochemical and structural formulas to complete the roster of possible rearrangements, we can summarize the transformations in the mechanistic diagram of Fig. 3. Each full line represents an overall suprafacial rearrangement, regardless of whether it occurs concertedly or stepwise via a biradical. In the latter case, biradical intermediates are to be imagined as lying along the pathway. Each dashed line represents an overall antarafacial rearrangement, concerted or stepwise. The relationship of Scheme 32 to this diagram can be seen by starting at the lowest vertex of Fig. 3, which represents the acetylene RZ-**80**. Note that RZ-**80** → SE-**81** is a suprafacial reaction, but RZ-**80** → RE-**81** is antarafacial, so that if both mechanisms were to run concurrently, the E-**81** allene product would be partially or fully racemized. On the other hand, if the reaction were mechanistically homogeneous (all suprafacial or all antarafacial), the E-**81** allene would be formed and the original acetylene would be recovered without loss of enantiomeric purity. Inspection shows that no mechanistically homogeneous pathway of whatever number of steps exists that permits loss of enantiomeric purity. However, racemization could occur by the mechanistically heterogeneous concurrence of the suprafacial and antarafacial pathways. These would be a direct consequence of the conformationally interconverting biradicals of Scheme 32 and would correspond in Fig. 3 to crossover between the full- and dashed-line pathways by way of interconversion of the biradicals.

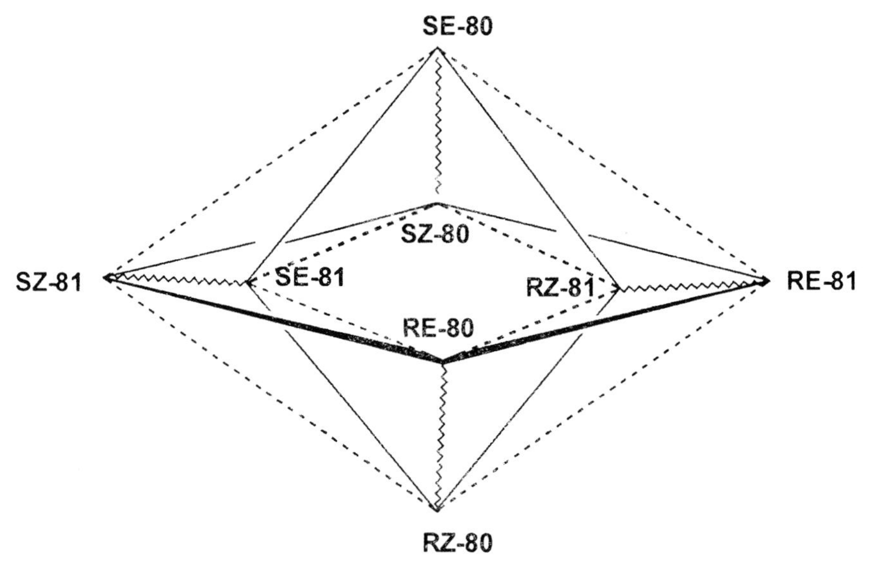

Figure 3. Interconversion paths among acetylenes **80** and allenes **81**. The stereochemical descriptors are indicated before the numerals. Solid and dashed lines, respectively, represent orbital symmetry allowed (suprafacial) and forbidden (antarafacial) sigmatropic rearrangements; wiggly lines represent stereomutation resulting from the equivalent of torsion about the olefinic bond by the mechanism of Scheme 32. Crossover between adjacent allowed and forbidden pathways also can occur by way of hypothetical interconverting biradicals, which are shown in Scheme 32.

Fig. 3 shows numerous paths interconverting the isomeric acetylenes and allenes. It will be instructive to follow one of these, chosen more or less arbitrarily. A simple pathway for the enantiomerization of the acetylene RE-**80** would be the sequence RE-**80** → SE-**81** → SE-**80**. This could occur in the biradical mechanism by the steps RE-**80** → **82b** → **82a** → SE-**81** → **82c′** → SE-**80** (Path **A up** in Table 1), where the biradicals **82b** and **82a** are the same as those bearing the same numbers in Scheme 32, and the biradical **82c′** is the enantiomer of **82c** of Scheme 32. Another possible path would be RE-**80** → RE-**81** → SE-**80**, which could occur in the biradical mechanism by the steps RE-**80** → **82c** → RE-**81** → **82a′** → **82b′** → SE-**80** (Path **B up** in Table 1).

Table I. Two Enantiomeric Paths for Interconversion of Acetylene Enantiomers by a Stepwise Biradical Mechanism of Cope Rearrangement

Path A down	Path A up	Path B down	Path B up
SE-**80**	RE-**80**	SE-**80**	RE-**80**
82c′	**82b**	**82b′**	**82c**
SE-**81**	**82a**	**82a′**	RE-**81**
82a	SE-**81**	RE-**81**	**82a′**
82b	**82c′**	**82c**	**82b′**
RE-**80**	SE-**80**	RE-**80**	SE-**80**

In discussing these reactions, we adopt the notation of Salem,[61] as we show in Figure 4, by projecting the profiles of Path **A** and path **B** onto the basal plane of the energy surface.

Note that although the trajectory of Path **A up** and its own step-by-step reverse (path **A down**) pass over the same intermediates, the sequence in which the intermediates are encountered is not the same, that is, the pathway does not possess reflection symmetry. Similarly, path **B up** is not the detailed reverse of path **B down**. Nevertheless, Fig. 4 conforms overall to the requirements of microscopic reversibility because each of the four paths of Table 1 has an enantiomeric partner, which is not its own step-by-step reverse. Thus, path **A down** and path **B up** are enantiomeric with each other, as are paths **A up** and **B down**. Enantiomeric points on the paired pathways are connected by the dashed lines in Fig. 4. In fact every point on the path **A** branch of Fig. 4 can be connected to an enantiomeric point on the path **B** branch by extending a straight line drawn from the point through the intersection point in the center of the diagram. It should be clear that for a mechanism

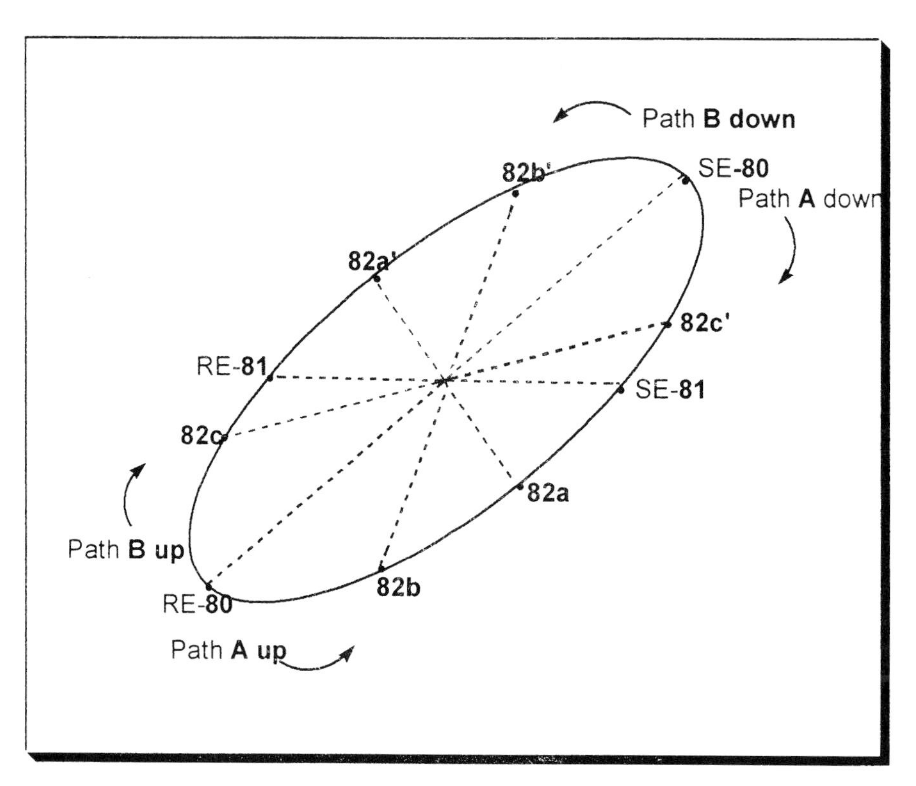

Figure 4. Schematic trajectories of interconversions of enantiomeric pairs excerpted from Fig. 3 and Table 1. The paths represent reaction coordinate changes starting at RE-**80** and ending at SE-**80**. Relative energy changes are to be imagined in the direction above the plane.

that interconverts enantiomers by paths lacking reflection symmetry, the energy surface can still satisfy pmr by having a C_2 rotation axis which interconverts enantiomeric structures.

Since all of the biradicals are chiral, and since, in fact, there is no point on any of the reaction itineraries that is achiral, the diagram of Fig. 3 represents in principle a way to manufacture racemic products out of enantiomerically enriched reactants with a machine that has no achiral components. In practice, this experimental design has been carried out for the **80–81** series, with the result that the rearrangements occur with complete preservation of enantiomeric purity.[62a–c] The rearrangement thus is facially homogeneous. Formally, this could be a consequence of either a completely concerted mechanism or of a biradical intermediate that undergoes bond cleavage at a rate much faster than conformational isomerization. Evidence given elsewhere[62a–c] suggests that the first of these possibilities is the correct one. In a study of a similar system modified to generate a more stable biradical, some racemization is observed, which indicates rearrangement with mixed faciality.[62d,e] The Cope rearrangement is thus shown to have a variable mechanism depending on the structure of the substrate.

5.12 The Direct Nucleophilic Displacement Reaction, and how we came to know its Stereochemistry

I have no doubt that if you were to try the following experiment, you would get the same answer that I did: Ask a random sample of organic chemists how they know that the direct nucleophilic displacement (S_N2) reaction occurs with complete inversion of stereochemical configuration, and the response invariably will cite the famous 1935 study by Hughes and co-workers[63] of the radioiodide exchange with 2-octyl iodide (**83–83′**, Scheme 33).

Scheme 33

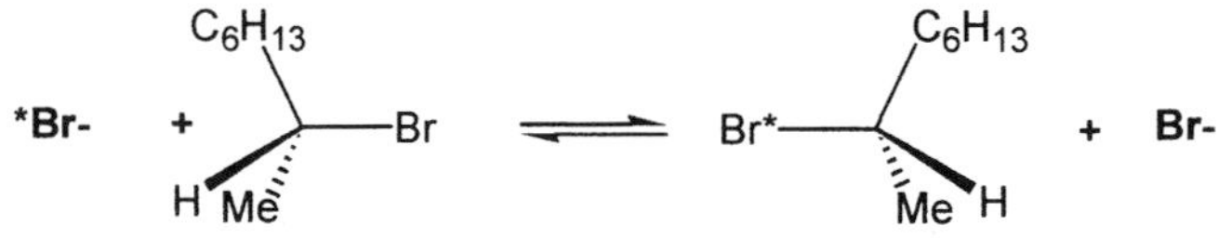

Br- + C_6H_{13}, H, Me –Br ⇌ Br– C_6H_{13}, Me, H + Br-

83 **83'**

A succinct statement of the results is given in Hughes's summary:

» *By means of the radioactive isotope of iodine, the velocity in acetone solution of the substitution of iodine for iodine in sec.-octyl iodide has been measured, and also the velocity of racemisation of* d-sec.*-octyl iodide by sodium iodide under similar conditions. The absolute rates of the two processes are the same*[64] *within the experimental error of the measurements of the radioactivity (10%). The result confirms in the most direct way possible the causal connexion between aliphatic substitution and optical inversion, in reactions of this type.* «

Hughes cited no precedent for this finding in 1935 or even in 1950 at the Montpellier Colloquium,[65] where he outlined this work and related experiments on α-phenylethyl bromide and α-bromopropionic acid from his laboratory in the words

» *The rates of bimolecular replacement can be deduced on the assumption that every individual act of substitution gives inversion. If the rates of replacement by the two procedures are identical, the assumption made concerning the stereochemistry of the process must be correct.* «

This conclusion is surely one of the most significant insights in all of mechanistic chemistry. However, Hughes's way of claiming to have found, essentially *de novo*, the proof of inversion in nucleophilic substitution had the effect (probably not intentional, as we shall see) of cloaking in obscurity a very large body of prior work leading to similar conclusions by others. Pre-eminent among the earlier contributions were those of Phillips and Kenyon. Fifty years ago,[66] or even later,[67] that work was appreciatively recognized. As time passed, however, it faded from view. Now most undergraduate and graduate textbooks contain no reference to it and mention only the Hughes work as evidence for the stereochemical course of what came to be called nucleophilic displacement. An objective decision on the justice of this asymmetric assignment of credit is not easy. Nevertheless, whatever one's view of that issue, I believe that a pedagogical opportunity is missed by the neglect of the Phillips-Kenyon work. Why did it slip into the shadows, to survive only in the memories of a few aging teachers who learned it decades ago? From the historical point of view, the answer to this question has much to tell us about how scientific ideas become accepted. Because of the decisive role of symmetrization in the ultimate outcome, the story is directly relevant to the theme of this chapter.

5.12.1 Phillips and Kenyon

Originally, the main objective of the extensive work of Kenyon and his collaborators was given in the general title of the series of dozens of papers: *Investigations on the Dependence of Rotatory Power on Chemical Constitution*. The project began in 1911 at the Municipal Technical School of Blackburn, north of Manchester, under the leadership of R.H. Pickard, with Kenyon as a close associate. In 1920, when Kenyon (Figure 5) moved to Battersea Polytechnic in London as Head of Department, he took the problem with him and gradually took its leadership over from Pickard, who had been made Principal of the Polytechnic in the same year.[68] At Battersea, the group's attention seems at first to have remained focused on the previous objective, which was of great interest to both organic chemists and theoreticians at the time.[69]

Figure 5. Joseph Kenyon, F.R.S. (1885–1961), Professor of Chemistry and Head of Department, Battersea Polytechnic. The image has been reproduced from ref. 68 by permission of the Royal Society.

Much confusion existed in the literature, because the widespread occurrence of Walden inversions cast doubt on attempts to correlate the configurations and rotations of a series of compounds. Kenyon recognized that a major step toward reliable assignments could be taken if one could find a way to assert with confidence that a specific reaction in a correlation occurred with inversion or with retention of configuration and whether the enantiomeric purity was fully conserved or partially lost. Henry Phillips, a junior associate of Pickard and Kenyon, took up work aimed at a clarification of the Walden inversion. In the end, the significance of Phillips's work transcended the original problem. His paper[70] on the question became a landmark in the development of mechanistic ideas. It is to Kenyon's credit that he apparently permitted Phillips to develop the methods to be described in 5.12.3 on his own, and moreover, that in a subsequent paper,[71] he freely acknowledged the younger man's sole responsibility for the important advance embodied in it.

A relevant evaluation of Kenyon's personality is given by Turner:[68]

»*As the Head of a chemical school which under his influence (one hesitates at the more formal word 'direction') became second to none as a source of fundamental and inspiring ideas, Kenyon was a most modest and at the same time compelling person ... He was, one might think, fortunate in his research collaborators: if so, it was surely his due; his work appealed to the intellectual rather than to the specialized or technical mind. He appears to have been a bit of a martinet at times when accuracy was in question and rightly so; but he was just withal.*«

5.12.2 The Walden Inversion

One cannot understand the history of these ideas without an appreciation of the Walden inversion. Kenyon's brief introduction to his lecture at the Montpellier Colloquium[72] caught the essence of Walden's discovery:

»*Over 30 years ago in connexion with the discovery of radioactivity, Sir Oliver Lodge said 'A discovery of real and essential novelty can never be made by following up a train of prediction. It is often made during the process of following a clue, but the clue does not logically lead to it ... The discovery which has been pointed to by theory is always one of profound interest and importance, but it is usually the close and crown of a long and fruitful period, whereas the discovery which comes as a puzzle and surprise usually marks a fresh epoch and opens a new chapter in science.'*

In his presidential address to the English Chemical Society in 1913, Professor Frankland remarked 'Walden's discovery was certainly a puzzle and a surprise, for it did not fit into any pre-existing theory of optical activity, and it had not been anticipated as a corollary to van't Hoff's theorem of the asymmetric carbon atom. It is highly probable that it may mark an epoch in our views with regard to the mechanism of the process of substitution in general.'«

Similarly, Hughes[65] summarized Walden's findings:

» *The question of the stereochemistry of substitution reactions is of special interest. The early concept concerning this problem, namely the naive assumption that, in a reaction of the form $Y + R\text{–}X \rightarrow Y\text{–}R + X$, the entering atom or group (Y) would replace the extruded entity (X) directly,* i.e., *without configurational change, was rudely shaken by the classical work of of Walden and others.*[73] *For the purpose of demonstrating the existence of the phenomena of the Walden inversion, two main methods have been employed. The first involves the conversion of an optically active compound into its enantiomorph by means of two or more reactions (Scheme 34). The second depends on the conversion of an optically active compound into enantiomorphous derivatives by the use of different reagents* « (Scheme 35).

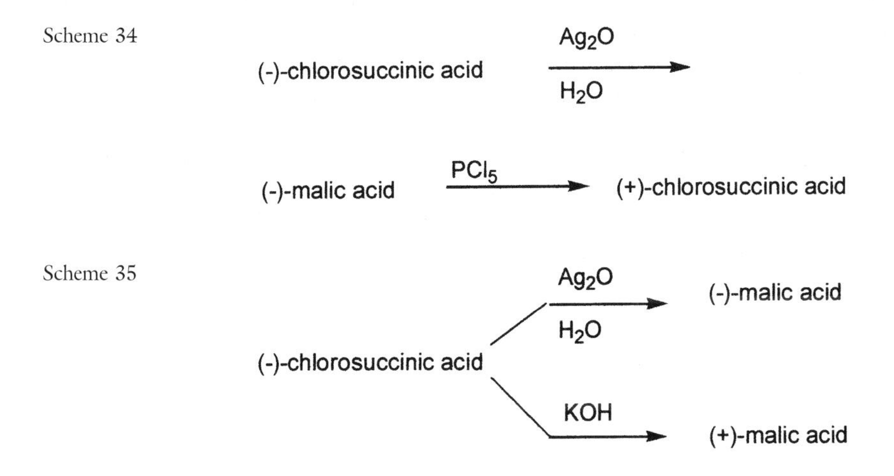

Phillips[70] succinctly outlined the uncertainty in the early interpretation of these Walden cycles:

» *The above scheme, due to Walden,*[73] *illustrates both types [of demonstrations of overall inversion of configuration]. They are, therefore, processes occurring in two stages. At one of these stages, an 'abnormal' reaction occurs resulting in a change of configuration, but at which one it is impossible to decide. In both reactions a group attached to the asymmetric carbon is replaced. Does the entering group in the first displacement take up the same position relative to the three remaining groups as that vacated by the group displaced? The sign of rotation of the product cannot be relied upon to give a correct answer to this question. Many instances are known in which substitution occurs remote from the asymmetric carbon atom, yet produces a reversal of sign. In such cases no change of configuration can be assumed, and therefore it is evident that change of sign does not necessarily coincide with change of configuration.* «

Hammett described the eventual achievement by the Kenyon-Phillips group of a sound basis for clarification:[63]

»*Experimental evidence has, however, reached a point where the principle that every displacement inverts is established with a high degree of reliability. This conclusion or indeed any definite one does not immediately emerge from the enormous amount of complicated and confusing data available. In addition to the fact that an actual cycle involves an unknown number of displacements, the racemization that frequently accompanies these reactions suggests the possibility of parallel reaction paths by which a reactant of one configuration leads to a product containing both stereoisomers. Finally, there is the unfortunate accident that the most easily available optically active compounds are the sugars, the terpenes, and the substances easily obtainable from amino acids or from lactic and malic acids. For reasons that will appear shortly, the reactions of these substances are exceptionally complex and difficult to interpret. The failure to recognize them as special cases has contributed largely to an utter and complete confusion, which leads one almost to suspect that some investigators have wished to prove that the situation is too complicated to permit any solution rather than by simplification of the problem to proceed stepwise toward one.*

A first and most important step in the direction of the necessary isolation of the variables was made in a series of investigations by Kenyon, Phillips, and their co-workers. Recognizing the probability that the easily accessible substances might offer especially complicated material, they undertook and solved the difficult problem of preparing a series of optically active alcohols containing only one center of asymmetry. Working with these, they soon found a number of clear and unambiguous cases of the following sort:« (Scheme 36).

Scheme 36

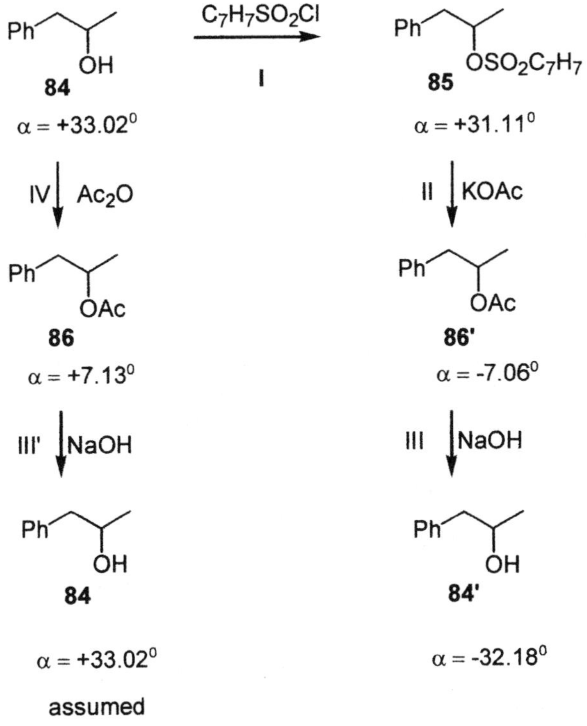

Ph–CH₂–CH(OH)–CH₃ **84** $\alpha = +33.02^{\circ}$ → ($C_7H_7SO_2Cl$, I) → Ph–CH₂–CH($OSO_2C_7H_7$)–CH₃ **85** $\alpha = +31.11^{\circ}$

84 → (IV, Ac_2O) → Ph–CH₂–CH(OAc)–CH₃ **86** $\alpha = +7.13^{\circ}$

85 → (II, KOAc) → Ph–CH₂–CH(OAc)–CH₃ **86'** $\alpha = -7.06^{\circ}$

86 → (III', NaOH) → Ph–CH₂–CH(OH)–CH₃ **84** $\alpha = +33.02^{\circ}$ assumed

86' → (III, NaOH) → Ph–CH₂–CH(OH)–CH₃ **84'** $\alpha = -32.18^{\circ}$

Scheme 36 is adapted from one given in the 1923 Phillips paper.[70] It demonstrates a clear case of Walden inversion in the transformation of the dextrorotatory 1-phenyl-2-propanol (benzyl methyl carbinol) **84** to its levoroatory enantiomer **84′**. Note that the rotations of the two alcohols are opposite in sign and, within experimental error, equal in magnitude. Relative to the earlier examples of Walden inversion, this cycle was uniquely informative, as Phillips had designed it to be:

» *In the examples of the Walden inversion to be described, however, we can on general grounds* single out the particular reactions which occur with configurational change *(emphasis supplied here). When d-benzylmethylcarbinol is allowed to react with* p-toluene-*sulphonyl chloride in the presence of pyridine, benzylmethylcarbinyl* p-*toluenesulphonate is obtained (see 84 → 85, Scheme 37).*

In this reaction, complete substitution of a group attached to the asymmetric carbon atom does not occur, the hydrogen atom of the hydroxyl group alone suffers displacement. It is justifiable therefore, to refer to the sulphonic ester prepared in this way from the dextrorotatory alcohol as the d-sulphonate ... «

» *The acetate prepared from the d-alcohol by the action of acetic anhydride is dextrorotatory(α+7.13°) and on hydrolysis with alcoholic sodium hydroxide yields the d-alcohol with unchanged rotatory power. That both these reactions should be accompanied by* complete *configurative change is unlikely and hence it may be assumed that dextrorotatory benzylmethylcarbinyl acetate has the same configuration as the dextrorotatory alcohol. In the above cycle of reactions (Scheme 36), however, this ester is obtained with a levorotation from a dextrorotatory sulphonate, the configuration of which can be said to be dextro.* «

Scheme 37

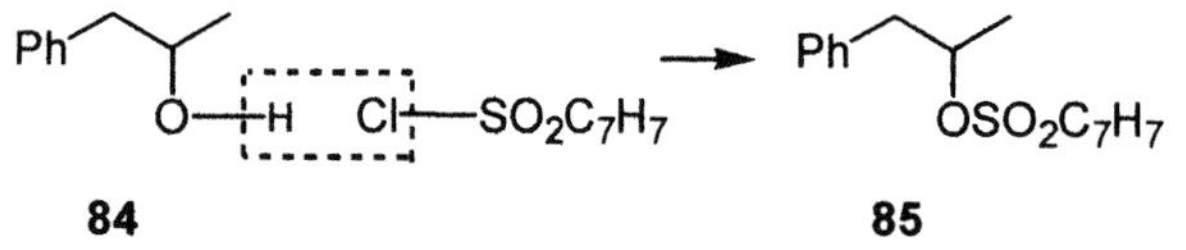

Ph, O—H Cl—$SO_2C_7H_7$ → Ph, $OSO_2C_7H_7$

84 **85**

Phillips concluded that the reaction shown as step II in Scheme 36 "therefore occurs by a definite abnormal reaction, and the almost complete inversion of configuration is clearly demonstrated by these reactions." He obtained similar results from the reaction of benzylmethylcarbinyl p-toluenesulfonate with potassium valerate: the benzylmethylcarbinyl valerate so formed gave back, upon alkaline hydrolysis, the carbinol with a rotation opposite in sign and unchanged in magnitude from that of the starting alcohol. Essentially, Phillips had swept away a whole area of uncertainty by relying upon a basic stereochemical principle, namely, given the rotation of a chiral compound, there is only other one compound whose rotation can be predicted from it with complete confidence: its enantiomer. Thus, although a transfomation of a molecule (+)-A to a different molecule (+)-B may correspond to either inversion or retention, transformation of (+)-A to its own enantiomer (-)-A must occur with inversion.

At this stage, the one point in the analysis that might have been questioned was Phillips's judgment that inversion in both the acetic anhydride acetylation of the carbinol and in the alkaline hydrolysis of the acetate were "unlikely." The nature of the problem can be seen in the two conceivable modes for the overall acetylation (Scheme 38).

Scheme 38

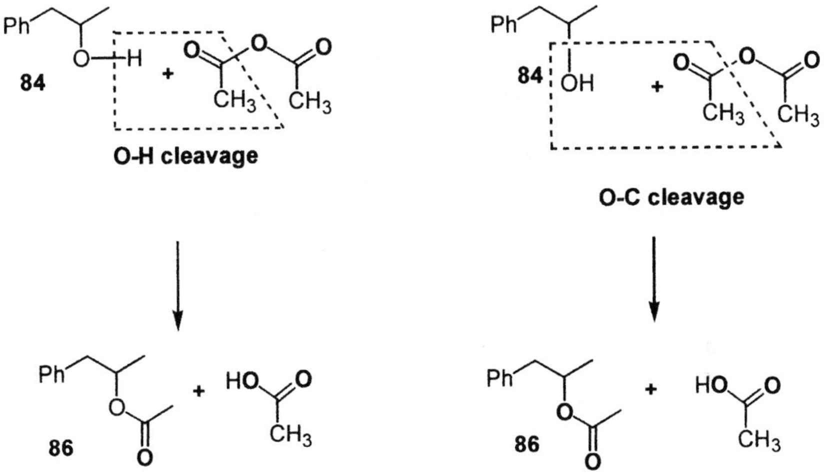

Ph, 84, O–H, +, O, O, O, CH_3, CH_3

O-H cleavage

Ph, 84, OH, +, O, O, O, CH_3, CH_3

O-C cleavage

Ph, 86, O, O, +, HO, O, CH_3

Ph, 86, O, O, +, HO, O, CH_3

If this process occurs by O–H cleavage (acyl transfer), application of the same argument Phillips had used to deduce retention in the tosylation would lead to the conclusion that acetylation completely retains configuration. However, if the acetylation occurs by O–C cleavage (acetoxyl transfer), a bond to the asymmetric center is broken during the reaction, and the possibility exists that partial or complete inversion could occur in this step. Similarly, the alkaline hydrolysis of the acetate could occur by O-acetyl cleavage, with assured retention of configuration, or by O-alkyl cleavage, with the risk of inversion.

Formal demonstrations of the direction of cleavage in alkaline hydrolysis of esters by the use of ^{18}O isotopic tracers would not be available for another eleven years.[74] Phillips gave no other supporting information from the literature; rather, he seemed to rely on the supposed improbability that *both* acetylation and alkaline hydrolysis of the acetate would have had to occur with inversion to invalidate his claim. Logically, in the absence of other evidence, if one admits that one of these processes can occur with inversion, there is no inherently greater probability that the other one can or cannot. The situation is analogous to that of a series of fair coin-tosses, in which the probability of "heads" on the second toss is one-half, independent of whether the first toss has led to "heads" or "tails."

In retrospect, it seems that it would have been simple to dispel the whole problem by making the acetate with acetyl chloride instead of acetic anhydride, in analogy to the strategy used for assuring retention in the *p*-toluenesulfonylation. Why Phillips did not adopt this approach is a puzzle.

Actually, there was already an example in the literature of this procedure. In 1912, Holmberg[75] had shown that acetylation of mandelic acid with acetyl chloride gave an acetate which, upon alkaline saponification, returned the original mandelic acid of identical sign and magnitude of rotation. Also, in the same year, Emil Fischer[76] had pointed out that although many cases of alkaline hydrolysis of sugar acetates were known, these never caused inversion of configuration. Since Fischer gave no indication of whether the acetates had been prepared with acetyl chloride or, as was common then, with acetic anhydride, the relevance of his comment to this issue is cloudy.

It is possible that Phillips knew of these results but did not cite them. More likely, in my opinion, he felt that the probability of O–C cleavage in acetylation of alcohols with acetic anhydride was low, by analogy in the then already well known acetic anhydride acetylation of amines, a reaction that gives only the N–H cleavage products (N-alkylacetamide, by acyl transfer, and acetic acid), not the N–C cleavage products (acetate, by acetoxyl transfer, and acetamide, Scheme 39).

Scheme 39

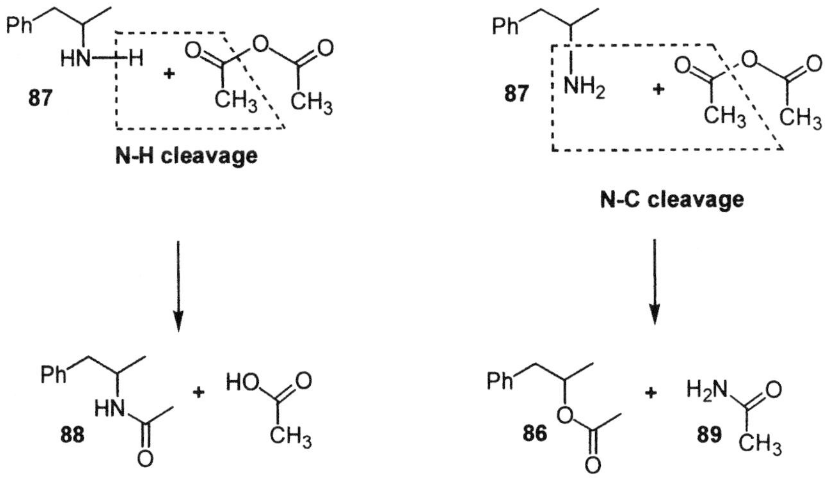

Ph HN–H 87 + O O O CH_3 CH_3
N-H cleavage

Ph NH_2 87 + O O O CH_3 CH_3
N-C cleavage

Ph HN O 88 + HO O CH_3

Ph O O 86 + H_2N O 89 CH_3

In due time, of course, the acyl chloride strategy was carried out in several other cases of acetylations of alcohols,[77,78] in which it was demonstrated that alkaline hydrolysis of the ester (acetate or benzoate) always gave back the original alcohol without loss of stereochemical integrity. Since ester formation from the acyl chloride does not break a bond to the asymmetric center, it must occur with retention, and therefore, the hydrolysis of the ester must also have occurred with retention. These results were consistent with the results of the later ^{18}O-labeling experiments,[74] which showed that the alkyl-oxygen bond in simple esters is not involved in alkaline hydrolysis (see Scheme 38). There was then little doubt that Phillips original assumptions on this issue had been correct.

By 1930,[69] several further cases of displacement of tosylates had been worked out by the Kenyon group using similar methods. These included those of ethyl lactate,[79] menthol,[80] and the *cis-* and *trans-* 2-methylcyclohexanols.[71] In 1940, Hammett[66] summarized the situation:

» *These results furnish a large and convincing body of evidence for the conclusion that the displacement of tolunensulfonate ion by a carboxylate or alkyloxy ion is invariably accompanied by inversion of configuration and suggest strongly that any nucleophilic displacement on carbon inverts.* «

How then did the chemical community come to lose sight of the origins of so significant a contribution? In my view, the reasons can be grouped into two major categories. The first of these had an unambiguously scientific basis, whereas the second grew from less easily recognized psychological and cultural causes.

5.12.3 The Significance of Kinetic Form

Up to this point, the experiments of Phillips, Kenyon and their co-workers had yielded a tightly reasoned although somewhat limited set of conclusions. They expanded the scope of their deductions slightly by means of the reasonable assumption that reactions of alkyl toluenesulfonates with other nucleophiles, such as halide ions, would proceed with inversion, by analogy to the reaction of the same substrates with carboxylates and alkoxides. However, they realized that in some nucleophilic substitutions, particularly those in which an α-phenyl substituent was present at the site of reaction, the predominant inversion reaction was accompanied by a certain amount of retention, leading to a partially racemized product.

The Kenyon group's attempts to explain this were misguided. They believed that the inversion reaction was "indirect" and took place through a carbonium ion intermediate.[69] The retention reaction, on the other hand, they believed to be "direct" and to result from exchange of the nucleophile and the departing ligand without alteration of the tetrahedral structure of the substrate.

A major deficiency in their reasoning was the failure to consider the consequences of their mechanistic proposals for the *kinetics* of the reactions. Their proposed "indirect" reaction, which involved two discrete steps, could give second-order kinetics only with a fast ionizing step followed by a slow rate-determining capture of the carbonium ion by the nucleophile. If the ionization were the slow step, the kinetics would be approximately first-order in substrate and zero-order in the nucleophile.

5.12.4 S_N1 and S_N2

It was not long until the Hughes-Ingold group at University College, London, showed that the Phillips-Kenyon mechanistic proposals were just reversed from the actual state of affairs. That story[65–67] is familiar to most chemists and may be summarized as follows: Nucleophilic substitutions at secondary carbon are cleanly second-order and occur with inversion via a bimolecular mechanism without an intermediate (S_N2), whereas certain tertiary and α-phenyl-substituted halides, under conditions such that the reverse reaction is suppressed, react by first-order kinetics via a unimolecular mechanism with an intermediate carbonium ion (S_N1). In the latter cases, partial racemization is common.

In spirit, the Hughes-Ingold experiments drew upon earlier studies on the mechanism of *electrophilic* substitution by Lapworth.[81] Hammett[82] refers to Lapworth's early work on the mechanism of bromination of acetone as "a classical example" of an investigation of mechanism which "initiated a new kinetic era in theoretical organic chemistry." Most organic chemists will recall that Lapworth detected the occurrence of an intermediate (the enol or enolate ion) in this reaction by his finding that the rate of bromination was independent of the concentration of bromine and depended on the concentrations of acetone and also of a catalyzing acid or base. It is clear that the Hughes-Ingold reasoning was conceptually parallel.

In fact, it would be a plausible conjecture that the design of the Hughes radiohalide exchange experiments themselves might have been stimulated by experiments on enolization in the immediately preceding years. Thus, in the period 1931–1933, Weissberger[83,84] used (and as far as I can determine, introduced) the technique of comparing an electrophilic substitution rate with a racemization rate in his elucidation of the mechanism of the autoxydation of benzoin and its derivatives. Similar studies on ketones[85,86] and sulfones[87] appeared soon after, so that one may safely assume that by the time of the Hughes radioexchange experiment in 1935,[63] this kind of comparison not only had become part of the *Zeitgeist* but actually was used by Ingold's group in Hughes's home department at University College. Hughes's application of a similar technique to nucleophilic substitution was of course ingenious, but it would not be surprising if the intellectual origins of that work included the immediately prior studies on electrophilic substitution.

The crucial importance of the Hughes radiohalide exchange experiments was not that they demonstrated that in the secondary derivatives, "every act of nucleophilic substitution occurs with inversion."[65] That already had been established by Phillips and Kenyon. Rather, Hughes's work showed that these reactions were bimolecular, and by analogy, suggested that inversion of configuration should be, with high probability, the result of any other bimolecular nucleophilic substitution.

In many papers and lectures,[65,88] Hughes and Ingold pounded away at the need to couple kinetic and stereochemical information in order to gain a firm understanding of the mechanisms responsible for the range of phenomena observed in the Walden inversion. Beyond this, however, they also stressed the need for such a coupling to make reliable correlations of structure, optical rotatory power, and configuration. This latter objective was the one Kenyon had been working toward for decades. I venture to speculate that Kenyon's long commitment kept him so tightly focused on this original program that he was unable to develop the broad insights needed to solve the problem of mechanism.

In the course of convincing an initially resistant chemical community, Hughes and Ingold sometimes felt the need to confront directly opposing positions held by other workers, among them Kenyon.[89] As we have seen, Hughes (Figure 6) could be outspoken in these discussions, but to judge from the portrait of his character given by Ingold,[89] it is unlikely that these disagreements should be interpreted as personally motivated. In fact, Hughes and Kenyon apparently were on friendly terms, as is implied in an anecdote told by Ingold:[89]

Figure 6. Edward David Hughes, F.R.S. (1906–1963), Professor of Chemistry, University College. The image has been reproduced from ref. 89 by permission of the Royal Society.

» *Never as a boy [in North Wales] did he [Hughes] disappear alone for the day into the wild and beautiful mountains at his doorstep. Never, indeed, in his life did he stand on a mountain summit save once, when, walking and talking chemistry with Joseph Kenyon in a summer mist on the track above Llanberis, the two of them, deeply absorbed and confined to the track by low visibility, reached the top of Snowdon by accident.* «

5.12.5 The Peculiar Attraction of Symmetrization Experiments

Phillips and Kenyon provided the first unequivocal method for identifying which step of a Walden substitution cycle was the inversion step. The Hughes-Ingold mechanistic insights directly depended upon that work, as Ingold himself[67] and others[66] certainly knew; in turn, they also provided a much firmer and more general basis for rational analysis of the stereochemical course of substitutions. Whether Phillips and Kenyon's failure to deduce the mechanism of the substitution justifies the eventual disappearance of their important contribution from the pedagogical mainstream is debatable, but one should keep in mind that, in science as in warfare, it is the winners who usually write the history of the conflict.

Sixty years later, it is possible to look behind the dazzle of the Hughes radiohalide experiments and realize that the isotopic exchange study was unavoidable in that work if one hoped to demonstrate that "every act of nucleophilic substitution occurs with inversion." As Hughes pointed out, without a knowledge of the rate of isotopic exchange, one could imagine, for example, that some of the iodide-for-iodide exchanges might take place by a retention mechanism, which in effect is not a reaction at all, since the product of such an exchange is indistinguishable from the reactant. In that case, the overall rate of enantiomerization would be slower than the rate of substitution. This problem, of course, does not arise in the Phillips-Kenyon work, where the substitution of an acetate for a toluenesulfonate is an irreversible process leading to a product other than the reactant. In fact, it would have been much simpler for Hughes to have studied the kinetics of the reaction of 2-bromooctane with potassium acetate. Had the result been second-order kinetics, the demonstration that in a bimolecular case, "every act of nucleophilic substitution occurs with inversion" would have been just as clearly demonstrated as in the radioiodide exchange experiments. The exchange experiments did have the advantage of showing unequivocally that halide-for-halide exchange, as had been previously assumed, followed the same stereochemistry as other nucleophilic substitutions, but as we have seen, that point was not a major issue by then.

But I think there is another set of reasons why the Kenyon-Phillips work has faded from view while the influence of the Hughes-Ingold contributions grew. First, although not conceptually unprecedented, as we have seen, the kinetic distinction between the broad mechanistic categories of nucleophilic substitution by Hughes and Ingold were carried out with an authoritative technical command that left little room for argument. The Hughes experiments in which the rates of radiohalide exchange and racemization were compared were especially impressive. They required techniques unfamiliar to most organic chemists, and the authors' apparent mastery of these procedures made the impact of the work all the stronger.

Beyond that, however, there remains the appeal of an exceptionally simple and recognizable symmetrization experiment. Although I have no quantitative survey data to substantiate this assertion, many years of association with organic chemists, both students and veteran professionals, convince me that this particular experiment, involving racemization via a symmetrical transition state or pathway, evokes a response of immediate assent and admiration.

5.13 The Human Response to the Symmetrical

We might ask: what property of symmetrization experiments accounts for the extraordinarily influential role they have played in research on reaction mechanisms? Partly it is the circumstance that two of the possible outcomes of such experiments, complete symmetrization or complete specificity, represent conceptual extrema. They have the appeal of all-or-nothing phenomena, whose interpretation calls forth the reductionist lurking in each of us. (Life is complicated enough, so we are happy when for once science is simple).

Perhaps this is all we need to know about the matter, but I wonder whether we are not struggling with a deeper thought, whose roots in the human psyche may be traced back at least as far as the beginnings of recorded history and perhaps much further. These roots seem to nourish a tree among whose branches we find both scientific and esthetic cognition. How can this be? Of course, the capacity of scientific events to elicit such responses has been widely remarked. One speaks of how "beautiful" a new molecule is, or of how "elegant" the solution to a mechanistic problem is. But can this impulse from the realm of feeling be understood at a deeper level?

The problem of relating scientific thought to the aesthetic and spiritual is similar to, if not actually overlapping with, one that emerges from considerations of the nature of entropy. It has been stated that[90]

» *In the physical sciences, the entropy associated with a situation is a measure of the degree of randomness, or of 'shuffled-ness' if you will, in the situation; and the tendency of physical systems to become less and less organized, to become more and more perfectly shuffled, is so basic that Eddington[91] argues that it is primarily this tendency which gives time its arrow – which would reveal to us, for example, whether a movie of the physical world is being run forward or backward.* «

Eddington himself remarks[91] that although

» *the law that entropy always increases – the second law of thermodynamics – holds, I think, the supreme position among the laws of Nature,* «

this pervasive concept of science has peculiar esthetic attributes:

» *Suppose that we were asked to arrange the following in two categories* – distance, mass, electric force, entropy, beauty, melody. *I think there are the strongest possible grounds for placing entropy alongside beauty and melody, and not with the first three. Entropy is only found when the* parts are viewed in association, *and it is by viewing or hearing the parts in association that beauty and melody are discerned. All three are* features of arrangement. *[emphasis added]. It is a pregnant thought that one of these three associates should be able to figure as a commonplace quantity of science. The reason why this stranger can pass itself off among the aborigines of the physical world is that it is able to speak their language, viz., the language of arithmetic.* «

In this passage, Eddington invokes the concept that it is the pattern of occupation of discrete quantum molecular energy levels by the members of a sufficiently large assembly of molecules that permits the entropy be calculated by statistical methods. Accordingly, the significance of the terms "parts in association" and "features of arrangement" becomes clear.

There, I think, may be one source of the powerful psychological impact of symmetrization experiments. Like entropy, beauty, and melody, symmetry is revealed only by viewing the parts in association. My application of Eddington's insight does not equate symmetry with beauty, of course, but it does ask: Because both symmetry and beauty are "features of arrangement," is it possible that a property common to both may generate the convictive power of such experiments? Is it possible that symmetry and symmetrization, being analyzable and hence recognizable, may serve purposes of broad importance in the functioning of biological organisms generally and of humans in particular, and that it is for these reasons that they elicit both utilitaritan and emotional responses that are relics of atavistic impulses? Are these merely vague speculations, or can they be examined in a systematic way?

If there is truth in such conjectures, the psychological and aesthetic significance of symmetry must have manifested itself long ago in history. We try to face this question now, beginning with a look at the influence of symmetry in certain aspects of early science. Then, at the end of the present chapter, we examine briefly some of what is known of the role of symmetry in modern cognitive and physiological sciences, and glimpse the much larger dimension of what is not known.

5.14 Symmetry in Cosmology and Aesthetics[92–95]

The aims of the concluding section of this chapter are to remind the reader that symmetry has exerted its attraction on the human spirit from early times and to understand, to whatever extent we can, the origin of this inclination. To begin, we follow the lead of Hermann Weyl, the mathematician whose brilliant little book[92a] on symmetry still retains its power to inspire scientists and aestheticians fifty years after its first appearance:

» *If I am not mistaken the word* symmetry *is used in our everyday language in two meanings. In the one sense symmetric means something like well-proportioned, well-balanced, and symmetry denotes that sort of concordance of several parts by which they integrate into a whole.* Beauty *is bound up with symmetry. Thus Polykleitos, who wrote a book on proportion and whom the ancients praised for the harmonious perfection of his sculptures, uses the word, and Dürer follows him in setting down a canon of proportions for the human figure. In this sense the idea is by no means restricted to spatial objects; the synonym "harmony" points more toward its acoustical and musical than its geometric applications.*

Ebenmass *is a good German equivalent for the Greek symmetry; for like this it carries also the connotation of "middle measure," the mean toward which the virtuous should strive in their actions according to Aristotle's Nicomachean Ethics, and which Galen in* De temperamentis *describes as that state of mind which is equally removed from both extremes.*

The image of the balance provides a natural link to the second sense in which the word symmetry is used in modern times: bilateral symmetry, *the symmetry of left and right, which is so conspicuous in the structure of the higher animals, especially the human body. Now this bilateral symmetry is a strictly geometric and, in contrast to the vague notion of symmetry discussed before, an absolutely precise concept.* «

We return later to this concept of "middle measure" in our consideration of the possible evolutionary pressures in the history of living organisms.

Weyl describes the symmetry operations of reflection and rotation, and then he relates these mathematical ideas to the spiritual concepts of the ancients:

» *Bilateral symmetry appears thus as the first case of a geometric concept of symmetry that refers to such operations as reflections or rotations. Because of their complete rotational symmetry, the circle in the plane, the sphere in space were considered by the Pythagoreans the most perfect geometric figures, and Aristotle ascribed spherical shape to the celestial bodies because any other would detract from their heavenly perfection.* «

The idealization of the sphere in ancient Greek thought went beyond these astronomical applications. Plato in *Symposium*[96] ascribed to the playwright Aristophanes an ingenious hypothesis, in the form of a creation myth, that was intended to explain the origins of both heterosexual and homosexual love. In this story, humankind originally included three sexes: male, female, and androgyne, derived respectively from the Sun, the Earth, and the Moon, "which is partly of the Sun and partly of the Earth." Each member of these sexes existed in a spherical shape, which was carried over from the shape of their body of origin. These creatures became quite powerful and threatened to attack the gods. To forestall this and to punish them, Zeus divided each entity in two, as a boiled egg could be divided by a hair, creating two males from each male, two females from each female, and a male and a female from each androgyne. In each divided individual, the only remaining visible trace of the originally spherical shape was the navel, which marked the place where Apollo, at Zeus's request, had stitched up the marks of the division. Aristophanes regarded love as the attempt of each half-human to heal the ancient wound of separation by finding the missing half and thus recapturing the lost unity. A modern evocation of this theme is found in Yeats's great poem *Among School Children:*[97]

» *... and it seemed that our two natures blent*
Into a sphere from youthful sympathy,
Or else, to alter Plato's parable,
Into the yolk and white of the one shell. «

In constructing his natural philosophy, Plato (427?–347 B.C.) drew upon the sphere as the exemplar of the divine. We must remember that his proposals were not based upon what we call scientific experimentation but were put forward axiomatically. In order for his ideas to be taken seriously as a plausible view of the essence of things, the axioms themselves already would have had to be so widely accepted at the time that others would consider them self-evident, and hence, the logical consequences derived from them reasonable. Because the creations of the gods would tend toward perfection, Plato incorporated the sphere into his scheme.

But a second requirement seemed to conflict with the notion that the ultimate constituents of matter were spherical particles. Leaping over much of Plato's argumentation, we nevertheless can recognize that any acceptable scheme of the nature of matter would have to take into account the evidence of the senses that matter takes different forms. It was therefore obvious to him that one would have to provide reasons in the underlying structure for the differences. His solution to this conundrum was clever. Whatever the shapes of the particles, they should conform as closely as possible to the spherical. It is here that the role of the the five regular polyhedra, tetrahedron, cube, octahedron, dodecahedron, and icosahedron emerges, for these are the only convex regular solids that can be inscribed in a sphere so that each corner of the solid just contacts the spherical surface, or in which a sphere can be inscribed so that it is tangent to the center of each face. In Plato's natural philosophy,[98] he associated the cube with earth, the tetrahedron with fire, the octahedron with air, and the icosahedron with water, while the dodecahedron stood for the universe as a whole.

By arguments too long to detail here,[98] Plato arrived at the proposal that these solids had a utilitarian purpose in the construction of the universe, that they in fact were the shapes of the ultimate constituents that, he imagined, when closely packed together, made up the world:

» *Before that time they (the four elements) were all without proportion or measure; fire, water, earth and air bore some traces of their proper nature, but were in the disorganized state to be expected of anything which god has not touched, and his first step when he set about reducing them to order was to give them a definite pattern of shape and number. We must thus assume as a principle in all we say that god brought them to a state of the greatest possible perfection, in which they were not before.* «

In this ancient thought, the ideas of symmetry and perfection are brought together.

5.15 Kepler, Symmetry, and Perfection

Centuries later, these same Platonic solids again took on a cosmological significance in Kepler's ingenious, though false, hypothesis for the ratios of the distances of the planetary bodies from the sun.[99] Kepler[100–102] (1571–1630) deserves our attention for several reasons: He "is universally esteemed as one of the major scientists of the seventeenth century and one of the greatest astronomers who has ever lived."[103a] He discovered

» *a new way of doing astronomy, which may be seen (at least in part) as a return to the authentic teaching of Plato in the* Timaeus *(in the sense of explanation in terms of both final and efficient causes), thereby effecting a revolution in method which has earned him the title of founder of modern astronomy.* « [103b]

Kepler worked at a crucial time of history, when the lingering night-fog of astrology and mysticism that clouded the thoughts of humankind was just beginning to be dissipated by the bright morning sun of science. But there is something else: Although Kepler was an astronomer rather than a chemist, his work on this problem embodies once more the age-old attraction for symmetry we have considered here; it reveals Kepler's own self-awareness of intellectual indebtedness to the Greek philosophers' idealization of symmetrical properties as controlling elements in the structure of the natural world; and it gives us an especially poignant demonstration that symmetry not only can serve as an organizing principle in science but also, that its seductive attraction, if not tempered with skepticism, can lead to delusion and error.

In Newman's evaluation,[101] Kepler

» *is usually considered the founder of physical astronomy. Copernicus conceived the heliocentric theory – reviving Pythagorean beliefs – and worked it out in his famous book* De Revolutionibus Orbium Coelestium; *Tycho Brahe invented and improved astronomical instruments, and by his wonderful skill in observation introduced undreamed-of accuracy into celestial measurements; Galileo contributed the telescope, the discovery of new stars and nebulae, the support and diffusion of Copernican ideas in his brilliant writings. Kepler ranks foremost as the mathematician of the sky … His fanatical search for simple mathematical harmonies in the physical universe produced some silly ideas but also three great laws. The first was that the planets move in ellipses with the sun in one focus. Before he made this discovery it was believed that the planets, being perfect creations of God, followed the most perfect of orbits, namely circles. [Compare the similar Aristotelian belief mentioned above that the heavenly orbs must be spheres]. The second law was that the line joining sun and planet (the radius vector) sweeps out equal areas in equal times. The third law was published in 1618, nine years after the other two. It connected the times and distances of the planets: 'The square of the time of revolution of each planet is proportional to the cube of its mean distance from the sun.'* «

5.16 Kepler and the Secret of the Universe

In 1596, when Kepler was a 24-year-old lecturer in mathematics at the University of Graz in the Austrian province of Styria, he published a report of his researches on the properties of the solar system, *Mysterium Cosmographicum*, now known in English as *The Secret of the Universe*. (One wonders how many other young scientists have dreamed of making on their own a discovery profound enough to deserve that title!). Kepler's own assessment of his findings can be summarized in a passage quoted by Lodge:[102]

» *The die is cast. I have written my book; it will be read either in the present age or by posterity, it matters not which; it may well await a reader, since God has waited six thousand years for an interpreter of his words.* «

Copernicus had put forward the heliocentric theory of the solar system in 1543, about 50 years before Kepler's *Mysterium*. Kepler found a challenging set of problems that emerged from the theory. Chief among these were: Why are there only six planets, and what physical property of the universe determines the ratios of the distances of the planets[104] from the sun?

For the sake of concision, we forego a trip through Kepler's tortuous reasoning, as detailed in the *Mysterium*, and proceed directly to his final proposal. In Kepler's theory, the major properties of the solar system (or as he called it, the universe) were a direct consequence of the influence of the Platonic bodies in Nature. As we have seen, these figures played a central role in Plato's theory of matter, a theory with which Kepler was thoroughly familiar. In fact Kepler may be said to have adapted Plato's vision of the microscopic components of the world to the large-scale problem of the existence and motion of the planets. Just as Plato had used the existence of five and only five regular solids to deduce the existence of the earth and the four elements, Kepler too deduced the existence of only six planets from the five spaces that would be created by inscribing the five solids within and around six spheres (Figure 7). The result that delighted Kepler most was his recognition from mathematical analysis that if one started with a sphere of unit radius representing the maximum solar distance in the orbit of Earth, circumscribed around it a dodecahedron, and circumscribed around that another sphere, the ratio of the radius of the second sphere to the first is the same as that of the minimum distance in the radius vector of Mars to that of the maximum distance of Earth. With great enthusiasm, he proceeded to try various combinations of the solids and their associated spheres until he found that the ratios of the radius vectors came into agreement with the experimentally determined ones when the outermost solid was a cube whose inscribed sphere marked the orbit of Saturn, the next was a tetrahedron for Jupiter, the next a dodecahedron for Mars, the next an icosahedron for Earth, the next an octahedron for Venus, in which was inscribed the sphere for Mercury. In Kepler's words,

» *The intense pleasure I have received from this discovery can never be told in words. I regretted no more the time wasted; I tired of no labour; I shunned no toil of reckoning, days and nights spent in calculation, until I could see whether my hypothesis would agree with the orbits of Copernicus, or whether my joy was to vanish into air.* «

Kepler's polyhedral hypothesis was an attempt to discern God's purpose to create the most beautiful and perfect world. Plato, in the *Timaeus*

» *emphasizes that, in explaining the origins of individual things, both mechanical causes and divine purposes must be considered, and moreover, if we wish to attain a true scientific explanation satisfying to human reason, we must be primarily concerned with the causes that lie outside the material in the realm of the spiritual. Aesthetic principles, such as those*

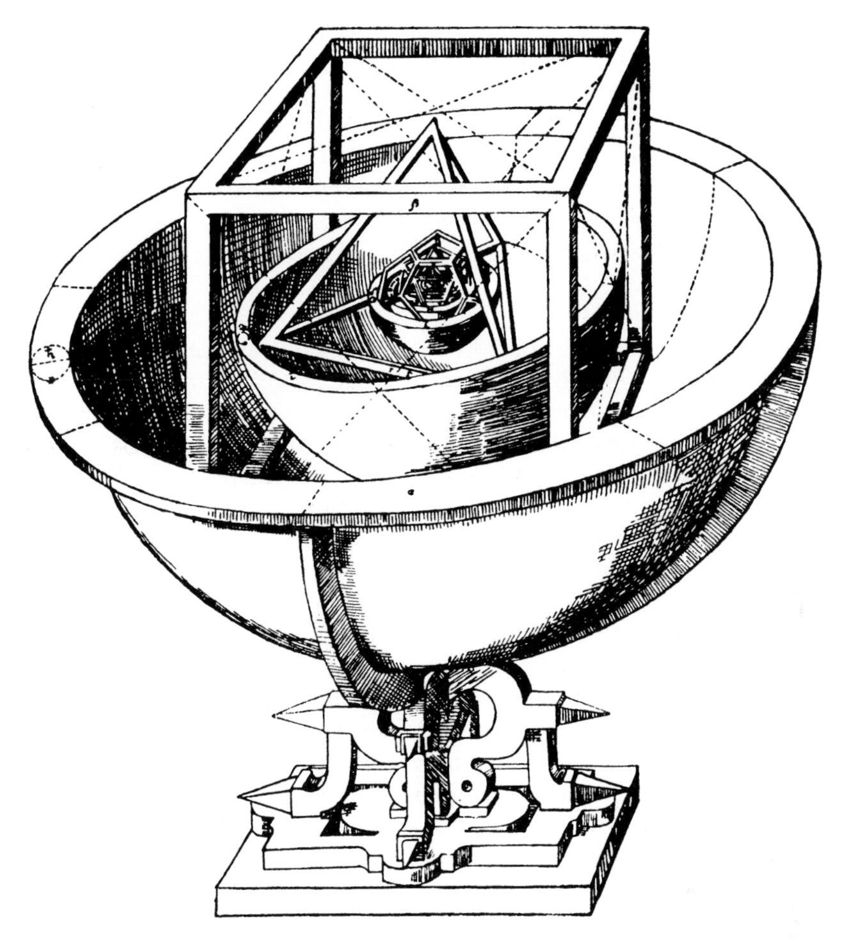

Figure 7. The five Platonic solids inscribed in concentric spheres whose radii are in the (approximate) ratios of the planetary distances, as imagined by Kepler. Taken from a copper engraving created for the 1596 edition of *Mysterium Cosmographicum*. Reproduced from the frontispiece of ref. 99.

of beauty and perfection, will serve as guides in the search for a priori causes; *for Kepler claims, quoting Cicero's translation of the* Timaeus, *that it was not possible for the perfect architect to create anything other than the most beautiful.* « [103c]

Accordingly, Kepler devoted substantial sections of *Mysterium Cosmographicum* to proposals of reasons why the Creator should have chosen that precise sequence of the planets to construct the solar system, speculations based largely upon a fancied association of the motions of each of the planets with the characteristic behavior pattern of the gods of classical mythology.

One therefore can understand readily how profound Kepler's joy at his discovery must have been. He believed that not only had he discovered the mechanism underlying the planetary distances, but also that he had succeeded in reading the mind and purpose of the Almighty. This urge to "know the mind of God" apparently still drives the work of some modern cosmologists.[105]

Unfortunately, however, it became clear centuries later that Kepler's "discovery" was purely imaginary and coincidental. First, the real distances of the six planets of his time later were found to be in only approximate agreement with those predicted by his theory, and more tellingly, additional planets, to a total of nine so far, are now known to be part of our solar system.

Caspar[106a] quotes a comment of Laplace[106b,107] made about Kepler some 200 years later:

» *It is depressing for the human spirit to have to see how this great man delightedly held up this chimerical speculation and thought it to be the soul and life of astronomy.* «

Caspar[106a] argues, and I would agree, that Laplace was too harsh on Kepler. My reason lies in the pattern that is evident to anyone who has been deeply involved in scientific research: Most theories are partial, incomplete, and provisional. Experience shows that it is unreasonable to expect them to survive for centuries before they are replaced by improved theories or by experimental facts. Of course, we recognize in Kepler's orbital radius hypothesis the residue of mediaeval and even ancient natural philosophy, but in the context of its time, it was a way of organizing what was (imperfectly) known about the solar system. It fulfilled the two major criteria for any theory: it explained (approximately) known phenomena, and it predicted something significant and ultimately testable, namely that the solar system should consist of six and only six planets. The aura of the mythical, mystical and supernatural that we now perceive to surround it was not so quaint and far-fetched at the time.

Laplace came to maturity, after all, in the era of the Enlightment, with its emphasis on the questioning of authority and on the empirical pursuit of science. I believe that what Laplace and other detractors reacted against was Kepler's inability to break away at one bound from the irrational forces still active in the immediate pre-scientific era. But in another sense, even later scientific advances followed the same track of hypothesis, experimental refutation, new hypothesis. Thus Kepler, in my view, differed most notably from later, more fully developed "scientific" scholars in the *nature of his ignorance*. The difference is important surely, but perhaps the impulse to dismiss this episode as an aberration should be tempered by the recognition that even a modern, ostensibly rational theory ultimately may be rejected because of the emergence of facts not available at the time of its promulgation.

Of course, we also must recognize that the well-spring of the erroneous theories of both Plato and Kepler was the belief in the dominant importance of symmetry in determining the shapes of the objects and phenomena of the natural world. Today, we understand that Nature is much richer, much more varied than humankind. The modern chemist, for example, need look no further than some of the molecules originally thought or assumed to be highly symmetrical: square planar cyclobutadiene, now known to be non-square,[108a,b] crown-shaped cyclooctatraene, now known to be tub-shaped,[108c,d] and regular hexagonal benzene triplet state, now known to be a distorted hexagon.[108e] The reasons why some molecules are symmetrical and some are not are buried deep within the binding forces of the structure, the comprehension of which has been a major concern of quantum mechanics. Whatever the underlying reasons for the attraction of *human beings* to the symmetrical, it seems clear that, given a choice, Nature follows no such general preference.

5.17 The Attraction of Symmetry: Biological and Evolutionary Necessity or Illusory Concept?[109–125]

It bears repeating that I use the terms "symmetry" and "symmetrization" in this discussion of chemical mechanisms to refer to *a reaction pathway* in which structural or stereochemical specificity is lost. As we have seen, this may, but need not, involve an actual point on the energy surface corresponding to a molecular species that is symmetrical in the group theoretic sense. In the following exploration of the possible connections between our present subject and the property called "symmetry" by aestheticians, psychologists, and philosophers we shall find a similar imprecision of usage. I think this is justifiable in both cases, because the term serves to remind us that the underlying mathematically rigorous situation, in which there is strict adherence to classifiability according to the symmetry operations of a definable point group, can serve as a standard or idealized model.

Much research in experimental psychology has been devoted to attempts to determine the relationship between biological processes and behavior. Issues in this field relevant to our present inquiry include the physiological basis of aesthetics,[94,95] the nature and purpose of the ability to perceive symmetry, a trait known to be strong in humans but also suspected to exist in other organisms,[109–112,118] the attempts to identify actual biological structures in the brain that underlie physiological responses,[110] and the evolutionary forces that shaped certain selective advantages conserved by the species for survival value.[111–113]

5.18 Gestalt Psychology and Aesthetics

Berlyne[94] points out that around the beginning of the 20th century, attempts to understand the psychological basis of aesthetics took two divergent forms. One was based upon psychoanalytic theories and concentrated on the analysis of individual artists and individual works. Some notable early examples include Freud's papers on Leonardo da Vinci and on Michaelangelo's *Moses*. In Berlyne's words,

» *Freud's explanation of the creation and appreciation of art had much in common with his explanation of other imaginative products, including dreams, slips of the tongue, jokes, fairy tales, neurotic symptoms, and psychotic fantasies. Like these other phenomena, art is interpreted as a vehicle of disguised expression for unfulfilled wishes, which are to a large extent unconscious … The ego and superego are usually responsible for frustrating direct gratification of the desires that must seek symbolic fulfillment as a substitute.* «[114]

The second approach comes from the work of the so-called "Gestalt" psychologists, who developed the concept of "goodness of configuration." According to them, we do not see isolated visual elements but rather "configurations" or "patterns" (*Gestalten*), which depend upon processes of perceptual organization occurring in the nervous system. In the *Gestalt* view,

» *some forms of organization are 'better' than others, and the brain tends to gravitate from 'worse' to 'better' configurations in accordance with the 'law of Prägnanz.' This law often has adaptive consequences. For example, the search for something that will round off a pattern and leave it with a 'good' structure can steer us toward appropriate actions or toward solutions of problems that we are thinking about.* « [115]

The properties which the *Gestalt* school associated most closely with "goodness" were regularity, symmetry, and simplicity. This idea has been criticized, not only because the term "good" is undefined but also because

» *although 'goodness' obviously exists in varying degrees, we are offered no guidance towards devising ways of measuring this property. Regularity, symmetry, and simplicity are readily recognizable as factors that play a large part in determining how satisfying a work of art will be. Here especially we badly need ways of differentiating the degrees to which these properties exist, or, in other words, of measuring them, since as we continually have occasion to note, the success of a work of art requires them to be present to just the right extent. A shade too much or a shade too little can often spoil things.* « [116]

Gombrich[95a] discusses the historical significance of the *Gestalt* view of perception as the first attempt, now largely superseded, to provide a detailed alternative to what came to be called the "bucket theory" of the mind. John Locke (1632–1704), the empiricist philosopher, had postulated that the mind of a new-born baby must be viewed as a *tabula rasa*, an empty slate; nothing could enter this mind except through the sense organs. Locke believed that only when these "sense impressions" became associated in the mind could we build up a picture of the world outside; there are no "innate ideas," man has no teacher except experience.[117]

It was Kant, a century later, who challenged this idea with the question: how can the mind ever order such impressions in space and time if space and time had first to be learned from experience? Without a pre-existent framework or "filing system" we could not experience the world, let alone survive in it. The idea of inborn reactions for which living organisms are "programmed" received much impetus from Darwin's work on evolution and more recently from studies in the field of ethology. If the idea is correct, it presents a question that is, to a considerable extent, the generative source of the discipline now known as physiological psychology or psychobiology. The objective of the work in this field is to identify the elements of the innate "filing system" or cognitive map by which humans and other organisms perceive the world.

In the context of this chapter, we may ask what is known today about the role of symmetry in this filing system and hence about the underlying mechanisms of our interaction with our environment. Since the literature is large and rapidly accumulating, we must select from among many studies only a few of special relevance here, without claiming to identify the best or most representative work in the field. In these articles, the authors attempt to answer such questions as: Is there an inherent "preference" for symmetry per se, or can behavior that seems to suggest such a preference be explained by alternative means? Whether or not symmetry is "preferred" in some aesthetic sense, do living organisms use symmetry or its absence as a guide to their response to the environment? What structures or mechanisms in the brain (or, in lower species, other location of sense) control the ability of the organism to recognize and use symmetry? If such structures exist, how did they come to prominence under evolutionary pressures? Of course, present-day scholars are approaching these questions in a much more focused and reductionist way than the *Gestalt* theorists did. Correspondingly, the answers are more tentative, more limited in scope, and one hopes, closer to reality.

5.19 Symmetry in Ethology and Psychobiology. Is There a Biological Preference for Symmetry?

This question has been addressed for species covering a range of the biological spectrum. A concise summary[111] of the findings of many authors is given in a paper by Enquist and Arak entitled *Symmetry, Beauty and Evolution:*

» *Humans and certain other species find symmetrical patterns more attractive than asymmetrical ones. These preferences may appear in response to biological signals, or in situations where there is no obvious signalling context, such as exploratory behavior and human aesthetic response to pattern. It has been proposed that preferences for symmetry have evolved in animals because the degree of symmetry indicates the signaller's quality. By contrast, we show here that symmetry preferences may arise as a by-product of the need to recognize objects irrespective of their position and orientation in the visual field. The existence of sensory biases for symmetry may have been exploited independently by natural selection acting on biological signals and by human artistic innovation. This may account for the observed convergence on symmetrical forms in nature and decorative art.* «

The preference for symmetry is not uniform in the arts, of course, as we have seen in Berlyne's analysis[94] and as our familiarity with the work of artists across the millenia makes clear. Even in nature, there are variations. With fruit flies,for example,[109b] attempts have been made to determine whether females have a sexual preference for large males with symmetrical characteristics of the sternopleural bristles and of sex comb teeth. The researchers compared the sizes and symmetry properties of copulating and non-copulating *Drosophila* males in these traits but found no differences. In their words, "these observations suggest that generalizations that large body size and symmetry promote mating success are unfounded."

On the other hand, certain birds apparently do show preferences for symmetry of various properties. Thus, females of the bluethroat species (*Luscinia s. svecica*) were shown[118] to associate more with males having symmetric rather than asymmetric leg bands.

For humans, a species much more intensively examined, we find some inconsistency. For example, some psychological experiments seem to indicate a preference by the observer for symmetrical faces, but more recent work suggests[109,119–121] that ascribing the cause of this observation to symmetry per se probably is an over-simplification. One study[121] concludes that "facial attractiveness is more dependent on physiognomy (of the owner) and less dependent on an asymmetrical perceptual process (in the observer) than is facial identity." Similarly,[109] a study of visual preferences of human infants for faces that varied in their attractiveness and in their symmetry about the midline was carried out with the aim of establishing whether infants' visual preference for attractive faces may be mediated by the vertical symmetry of the face. The result:

» *Chimeric faces, made from photographs of attractive and unattractive female faces, were produced by computer graphics. Babies looked longer at normal and at chimeric faces than at normal and unattractive faces. There were no developmental differences between the younger and older infants: all preferred to look at the attractive faces. Infants as young as 4 months showed similarity with adults in the 'aesthetic perception' of attractiveness and this preference was not based on vertical symmetry of the face.* «

Another study[119] pointed out that previous investigations of the relationship between asymmetry and facial attractiveness have confounded manipulations of asymmetry with facial "averageness" and mean trait size. The authors then performed a manipulation that altered asymmetry within a face without altering the mean size of facial features. Contrary to what was predicted, faces that were made more symmetrical were perceived as less attractive. They concluded that "these results do not support the hypothesis that attractiveness is related to [high levels of symmetry.]" Similarly, it has been suggested[119] that the females' frequently observed mating choice of symmetrical males among certain species may not arise from a preference for symmetry as such. Instead, it may be simply a consequence of the fact that the characteristics of symmetrical males are closer to the species population average. (Recall our earlier discussion in Section 5.14 of the concept of "middle measure" as having a moral and ethical significance in Greek philosophy). From an evolutionary point of view, it seems reasonable that individuals conforming to the prevalent characteristics of a living and hence self-evidently viable population would be perceived as favorable to the production of healthy offspring. In that case, symmetry would be functioning as an indicator of the true quality of the male, namely averageness.

5.20 Visual Detection of Bilateral Symmetry

As we now have seen, it is difficult to single out symmetry from other confounding influences as the primary cause of aesthetic or behavioral responses in living organisms. Is it nevertheless possible to determine whether symmetry itself may play a role in more narrowly defined perceptual tasks? An approach to this kind of "reduced" problem is exemplified by a study of bilateral symmetry detection.[110]

It has been known for some time that bilateral symmetry is more easily recognized by humans when the plane of symmetry of the object being viewed is vertical with respect to the retina (parallel to the viewer's own symmetry plane) than when it is at other orientations. This idea was proposed as a hypothesis in the late 19th century by the remarkable Viennese polymath Ernst Mach.[122] As far as I have been able to determine, Mach offered this simply in categorical terms, without empirical evidence (other than perhaps that derived introspectively), but since then numerous experimental studies of symmetry detection in human subjects, reviewed by Herbert,[110b] support the idea as a "robust finding."

Mach suggested that it is the bilateral symmetry of the ocular musculature that underlies the vertical advantage. Julesz[123] and a number of later workers expanded on this idea in efforts to determine more specifically the region of the brain actually responsible. Julesz noted that the projections of a vertically symmetric pattern onto the visual system would result in the left half of the image *first* going to the right cortical hemisphere, and the right half *first* going to the left hemisphere. He suggested that some point-by-point matching process occurs between symmetrically opposite loci in each cortical hemisphere, and others[124] noted that this matching process could be mediated by fibres crossing over through the corpus callosum.[123]

Thus, the neuroanatomy of the human visual system could support the point-to-point matching required by the callosal hypothesis. These considerations lead to the predictions that the detectability of symmetry should be narrowly tuned around vertical, and that presentation to a test subject of patterns away from this orientation should decrease the accuracy with which bilateral symmetry can be detected. They also predict that subjects born without a corpus callosum should not detect vertical symmetry more easily than that at other orientations. Recent experimental evidence[110] seems to be in good accord with these predictions.

A fascinating development in the relationship between symmetry and esthetics has emerged recently from a study by Tyler[125] of artists' portraits (oil paintings, watercolors, drawings and engravings) of single persons in which both eyes are visible. The data examined included work of 265 artists over a 600-year period, representing the careers of old masters such as van der Weyden, Botticelli, Leonardo, Titian, Rubens, and Rembrandt, as well as many 20th century painters. Tyler[125] points out that:

» *The importance of the centre of the canvas has long been appreciated in art, as has the importance of the eyes in revealing the personality of subjects of portraits. The center of symmetry of the face is often discussed in art analysis and might be expected to be used as an explicit principle of composition by artists trained according to such analysis. However, I have found that portraits painted throughout the past 600 years adhere to a different compositional principle not discussed in the literature: one eye [the forward one] is consistently centred horizontally in the canvas.* «

The distribution of locations of the mouth or of the facial midline of bilateral symmetry is not peaked in this manner. Tyler concludes

» *My analysis shows that explicit compositional principles are implemented with an unbiased accuracy of ±5% over the past six centuries. This precision results from perceptual processes that seem to be unexpressed by the artists themselves, suggesting that hidden principles are operating in our aesthetic judgements, and perhaps in many realms beyond portraiture.* «

Although overall symmetry of images cannot be said to be uniformly favored in art, Tyler's work apparently has brought to light one element of symmetry that has lurked in the artistic unconscious for centuries and has dominated the placement of facial images in the picture frame.

5.21 Conclusions

To this lay observer, the current status of research in aesthetics and in psychobiology does not provide a basis for the assertion that symmetry is the dominant influence on the perception of beauty or on physiological response to the environment. It is *one* influence, but except for a few narrowly defined functions, its role usually is mixed with that of other factors. Yet there is no doubt that symmetry strongly influences the intellectual and even emotional response of chemists (and, for all I know, of other scientists) to experimental information and of artists to their depictions. I must confess that I have not been able to trace the ultimate origins of such responses. In this sense, Tyler's remark[125] that artists follow "hidden principles" in their work comes close to my present view of the matter. I conjecture that the origins are related to symmetry's function as an aid to recognition of objects by simplification. But I also wonder whether the influence of symmetry, in one or another of its manifestations, on artists, on the modern chemist, on the ancient Greek philosopher, and on Kepler and other early astronomers, may arise from deeper, more complex roots, not yet accessible to scientific study, roots from which spring humankind's aspiration toward the sublime. You will no doubt recognize this as an acknowledgment of frustrated purpose. I can only respond that the outcome of my examination of the subject seems to match that of Weyl,[92a] who wrote a similar statement of what may well be the same conjecture:

» *Symmetry, as wide or as narrow as you may define its meaning, is one idea by which man through the ages has tried to comprehend and create order, beauty, and perfection. It is in this tradition that a modern poet [Anna Wickham[92e]] addresses the Divine Being ...*

'God, Thou great symmetry,
Who put a biting lust in me
From whence my sorrows spring,
For all the frittered days
That I have spent in shapeless ways
Give me one perfect thing.' «

5.22 Acknowledgment

Two paragraphs of this chapter have been adapted from an earlier article: Berson, J.A. *Chemtracts – Organic Chemistry*, **1989,** 2, 213. D.M. Berson, K. Mislow, and P. Gay provided helpful advice and literature references.

5.23 References

(1) Slater, J.C. *Introduction to Chemical Physics,* 1st ed. McGraw-Hill, New York, NY, 1939, p. 120ff.
(2) Mason, S.F. *Nouveau J. Chim.*, **1986,** *10*, 739, and references cited therein.
(3) Beckmann, E. *Ann.* **1889,** *250*, 322.
(4) Semmler, F.W. *Die Ätherischen Öle*, Verlag von Veit & Comp. Vol. III, Leipzig, 1906, p. 284ff, and references cited therein.
(5) Berson, J.A. *Chemical Creativity: Ideas from the Work of Woodward, Hückel, Meerwein, and Others.* Wiley-VCH Publishers, New York and Weinheim, 1999, Chapter 4.
(6) Simonsen, J.L.; Owen, L.N. *The Terpenes*, 2nd ed. Vol. I, Cambridge University Press, Cambridge, 1947, p. 232–233.
(7) (a) Historical review: Ingold, C.K. *Structure and Mechanism in Organic Chemistry*, 2nd ed. Cornell University Press, Ithaca, 1969, p. 794ff. Natural products chemistry again drove part of this development, since much of Baeyer's work (ref. 7b) in the area centered on indoxyl and isatin, oxidation products of the ancient blue dye indigo, which is extracted in glycosidic form from various plants. (b) Baeyer, A. *Ber.* **1878,** *11*, 1296; **1879,** *12*, 456.
(8) Beckmann, E.; Mehrländer, H. *Ann.* **1896,** *289*, 362.
(9) Beckmann, E. *Ber.* **1886,** *19*, 988, and subsequent papers.
(10) Historical review: Meisenheimer, J.; Theilacker, W. in *Stereochemie*, Freudenberg, K. ed. Franz Deuticke, Leipzig, 1933, p. 964ff.
(11) (a) Wallach, O. *Ann.* **1893,** *276*, 296. (b) For the origin of Wallach's work on terpenes, see Chapter 4, ref. 4.
(12) Lowry, T.M. *J. Chem. Soc.* **1899,** *75*, 211.
(13) Lapworth, A. *J. Chem. Soc.* **1892,** *61*, 808.
(14) Lapworth, A.; Hann, A.C.O. *J. Chem. Soc.* **1902,** *81*, 1502.
(15) Lapworth, A.; Hann, A.C.O. *J. Chem. Soc.* **1903,** *83*, 111.
(16) Kipping, F.S.; Hunter, A.E. *J. Chem. Soc.* **1903,** *83*, 1009.
(17) Tutin, F.; Kipping, F.S. *J. Chem. Soc.* **1904,** *85*, 65.
(18) This could hardly have been an oversight, since in the same publication,[17] the Wallach paper is cited in another context.
(19) Wagner-Jauregg, T. ref. 10, p. 858.
(20) Willstätter, R. *Aus Meinem Leben*, Verlag Chemie, Weinheim, 1949, p. 3.
(21) Simonsen, J.L.; Owen, L.N. The Terpenes, 2nd ed. Vol. II, Cambridge University Press, Cambridge, 1949, Chapters IV and V.
(22) Berson, J.A. in *Molecular Rearrangements*, de Mayo, P. Ed. Interscience, New York, 1963, Volume I, Chapter 5.
(23) A quality variously translated as feeling, touch, or sensitivity.

(24) Wagner, G. *J. Russ. Soc. Phys. Chem.* **1899,** *31*, 680.
(25) (a) Fittig, R. *Ann.*, **1860,** *114*, 56. (b) For a review of the early literature, see Wallis, E.S. in *Organic Chemistry, An Advanced Treatise*, Gilman, H., ed. Wiley, New York, 1943, Vol. I, p. 968ff.
(26) (a) For a review of Meerwein's scientific work, see, Criegee, R. *Angew. Chemie. Intl. Ed. Engl.* **1966,** *5*, 333. (b) For a review of Meerwein's academic career, see Dimroth, K. *Angew. Chemie. Intl. Ed. Engl.* **1966,** *5*, 338.
(27) Meerwein, H.; Unkel, W. *Ann.* **1910,** *376*, 152.
(28) Meerwein, H.; Probst, H.; Mühlendyk, W. *Ann.* **1914,** *405*, 129.
(29) Erlenmeyer, E. *Ber.* **1881,** *14*, 322. Numerous later authors adopted this mechanism.
(30) Breuner, A.; Zincke, T. *Ann.* **1879,** *198*, 141.
(31) Meerwein, H.; Probst, H.; Kremers, F.; Splittegarb, R. *Ann.* **1913,** *396*, 200.
(32) Tiffeneau, M. *Rev. générale des Sci. pures et appls.* **1907,** *18*, 583.
(33) Nef, J.U. (a) *Ann.* **1892,** *270*, 267. (b) Review: *ibid.* **1897,** *298*, 202. (c) *J. Am. Chem. Soc.* **1904,** *26*, 1549.
(34) Michael, A. *Ber.* **1901,** *34*, 918.
(35) Stieglitz, J. *Amer. Chem. J.* (a) **1896,** *18*, 751. (b) **1903,** *29*, 49.
(36) Tiemann, F. *Ber.* **1891,** *24*, 4162.
(37) Purists might argue that the "environment" is not truly achiral except at complete conversion, whereas in practice during the reaction some unreacted chiral borneol remains; however, for this formal "chirality" to be effective in generating some optical activity in the camphene would have required that the kinetic order of the reaction be higher than unity in borneol. Meerwein either did not think of this possibility or simply ignored it as captious; later kinetic studies of Wagner rearrangements of various bornyl and isobornyl derivatives justified his position by showing strict first-order behavior.
(38) Semmler, F.W. *Ber.* **1902,** *35*, 1016
(39) Semmler, F.W. *Die Ätherischen Öle*, Verlag von Veit & Comp. Leipzig, 1906. (a) Vol. II, p. 73, and references cited therein. (b) *ibid.* Vol. III, p. 111.
(40) (a) Meerwein, H.; Fleischhauer, C. *Ann.* **1918,** *417*, 255. (b) Fleischhauer, C. *Inaug. Diss.*, Bonn, 1915, as cited in ref. 40a.
(41) Ruzicka, L. *Helv. Chim. Acta* **1918,** *1*, 110.
(42) Bredt, J.; Holz, W. *J. prakt. Chem. Neue Folge* **1917,** *95*, 133.
(43) In Scheme 13, which shows Meerwein's mechanism, the bridgehead alkene intermediate **35** would be considered unlikely today, even as a transient species, but at least its further reaction, hydration to camphene hydrate, is plausible. The corresponding bridgehead alkene intermediate from the divalent species **42** of Scheme 15 would have to undergo *anti-Markownikoff* hydration to continue on to an eventual stable Wagner product.
(44) In tracing the history of peer review in *Helvetica Chimica Acta*, I have been informed by Dr. M.V. Kisakurek, the present Editor in Chief of that journal, that the system of reviewing papers submitted for publication was first introduced around 1970. Prior to that, the decisions were made by the "omnipotent" president of the redaction committee, without any external help on the quality of the submitted manuscripts. I thank Dr. Kisakurek for kindly providing this information. Who the president of the redaction committee was in 1918 is not clear, since the *Helvetica* did not start listing the names of the committee membership until 1932.
(45) (a) Meerwein, H.; van Emster, K. *Ber.* **1920,** *53*, 1815. (b) Meerwein, H.; van Emster, K. *Ber.* **1922,** *55*, 2500.

(46) (a) Aside from a small rotation ascribed to the presence of trace contamination by camphene, the tricyclene sample was optically inactive, as expected. (b) Lipp, P. *Ber*. **1920,** *53*, 769, in a paper published independently just before that of Meerwein and van Emster, showed that tricyclene is unreactive under the "Bertram-Walbaum" conditions ($ZnCl_2$/HOAc) which convert camphene to isobornyl acetate.

(47) (a) Bartlett, P.D. in *Organic Chemistry, An Advanced Treatise*, Vol. III, Wiley, New York, NY, 1953, p. 66. (b) Ingold, C.K. *Structure and Mechanism in Organic Chemistry*, Cornell University Press, New York, NY, 1st ed. 1953, p. 482. (c) Nevell, T.P.; de Salas, E.; Wilson, C.L. *J. Chem. Soc*. **1939,** 1188. (d) Reviews *inter alia* in refs. 22 and 47e–i. (e) Bartlett, P.D. *Nonclassical Ions*, W.A. Benjamin, New York, NY, 1965. (f) Olah, G.; Schleyer, P. von R. *Carbonium Ions*, Vol. 3, Wiley, New York, NY, 1972. (g) Kirmse, W. *Top. Curr. Chem*. **1979,** *80*, 125. (h) Brown, H.C. (with commentary by Schleyer, P. von R.) *The Nonclassical Ion Problem*, Plenum Press, New York, NY, 1977. (i) Bartlett, P.D. *J. Am. Chem. Soc*. **1972,** *94*, 2161. (*The Scientific Work of Saul Winstein*, an appreciation, in the Winstein Memorial issue of *J. Am. Chem. Soc.*). (j) Olah, G. *Angew. Chem. Intl. Ed. Engl*. **1995,** *34*, 1393 (Nobel Prize Lecture), and references cited therein. (k) Saunders, M.; Kates, M.R. *J. Am. Chem. Soc*. **1980,** *102*, 6867. (l) Saunders, M.; Kates, M.R. *J. Am. Chem. Soc*. **1983,** *105*, 2889.

(48) (a) Meerwein, H.; Wortmann, R. *Ann*. **1924,** *435*, 190. (b) Although the authors were noncommittal on the actual stereochemical course of this proposed shift, I have shown it here as endo-to-endo, in accord with observations in a related case (ref. 48c). (c) Berson, J.A.; Grubb, P.W. *J. Am. Chem. Soc*. **1965,** *87*, 4016.

(49) Meerwein, H.; Montfort, F. *Ann*. **1924,** *435*, 207.

(50) (a) Houben, J.; Pfankuch, E. *Ann*. **1933,** *501*, 219. In ref. 22, this reference was inadvertently omitted and its contents were incorrectly ascribed to the present ref. 50c. (b) Nametkin, S.; Brüssoff, *Ann*. **1927,** *459*, 144. (c) Houben, J.; Pfankuch, E. *Ann*. **1931,** *489*, 193. (d) Bredt, J. *J. Prakt. Chem*. **1931,** [2]*131*, 144. (e) Houben, J.; Pfankuch, E. *Ann*. **1930,** *484*, 273. (f) Roberts, J.D.; Yancey, J.A. *J. Am. Chem. Soc*. **1953,** *75*, 3165. (g) Vaughan, W.R.; Perry, R., Jr. *J. Am. Chem. Soc*. **1953,** *75*, 3168. (h) Roberts, J.D.; Lee, C.C. *J. Am. Chem. Soc*. **1951,** *73*, 5009. (i) Doering, W. von E.; Wolf, A.P. *Perfumery Essent. Oil Record*, **1951,** *42*, 414.

(51) Loftfield, R.B. *J. Am. Chem. Soc*. (a) **1950,** *72*, 632. (b) **1951,** *73*, 4707.

(52) Stork, G.; Borowitz, I.J. *J. Am. Chem. Soc*. **1960,** *82*, 4307.

(53) House, H.O.; Gilmore, W.F. *J. Am. Chem. Soc*. **1961,** *83*, 3980.

(54) Aston, J.G.; Newkirk, J.D. *J. Am. Chem. Soc*. **1951,** *73*, 3900.

(55) Burr, J.G.; Dewar, M.J.S. *J. Chem. Soc*. **1954,** 1201.

(56) For a list of reviews, see March, J.A. *Advanced Organic Chemistry*, Wiley, New York, 4th ed. 1992, p.1081.

(57) (a) Cordes, M.H.J.; Berson, J.A. unpublished work. (b) Cordes, M.H.J. Ph. D. Dissertation, Yale University, 1994, p. 148 ff.

(58) (a) We disregard for the present purpose the issue of whether transition state theory is appropriate to any individual case. A discussion of reactions in which it may not be is given by: (b) Carpenter, B.K. *J. Am. Chem. Soc*. **1985,** *107*, 5730 and following papers.

(59) (a) Mislow, K. *Science*, **1954,** *120*, 232. (b) Mislow, K. *Introduction to Stereochemistry*, Benjamin, New York, 1966, p. 93. (c) Mislow, K. *Science*, **1950,** *112*, 26. For later discussion, see: (d) Mislow, K. *Accounts Chem. Res*. **1970,** *3*, 321.

(60) Burwell, R.L.; Pearson, R.G. *J. Phys. Chem*. **1966,** *70*, 300.

(61) Salem, L.; Durup. J.; Bergeron, G.; Cazes, D.; Chapuisat, X.; Kagan, H. *J. Am. Chem. Soc.* **1970,** *92*, 4472.
(62) (a) Owens, K.A.; Berson, J.A. *J. Am. Chem. Soc.* **1988,** *110*, 627. (b) Owens, K.A.; Berson, J.A. *J. Am. Chem. Soc.* **1990,** *112*, 5973. (c) Berson, J.A. *Chemtracts-Organic Chemistry* **1989,** *2*, 213. (d) Wessel, T.E.; Berson, J.A. *J. Am. Chem. Soc.* **1994,** *116*, 495. (e) See also: Duncan, J. A.; Hendricks, R. T.; Kwong, K. S. *J. Am. Chem. Soc*, **1990,** *112*, 8433.
(63) Hughes, E.D.; Juliusburger, F.; Masterman, S.; Topley, B.; Weiss, J. *J. Chem. Soc.* **1935,** *137*, 1525.
(64) Hughes defined the rate constant for substitution as the total rate constant in all four types of reaction ((+)- and (-)-2-octyl iodide with either $*I^-$ or I), and the rate constant for racemization was defined as the total rate constant for inversion in both directions. It is these rate constants that Hughes found to be equal. Another way to approach the problem is to define a rate constant k_1 for substitution in one direction. On the assumption that every substitution inverts, the rate constant for enantiomerization should also be k_1, but the rate constant for loss of optical activity would be $2k_1$.
(65) Hughes, E.D. *Bull. Soc. Chim. Fr.* **1951,** C17.
(66) (a) Hammett, L. *Physical Organic Chemistry*, McGraw-Hill, New York, 1940, p. 160. (b) ibid. p. 163.
(67) Ingold, ref. 7a, p. 512.
(68) Turner, E.E. Memoir of Joseph Kenyon, *Biogr. Mem. Fellows Roy. Soc.* **1962,** *8*, 49.
(69) See for example the symposium of the Faraday Society on optical rotatory power and constitution at which Kenyon was a participant: Kenyon, J.; Phillips, H. *Trans. Faraday. Soc.* **1930,** *451* (1930).
(70) Phillips, H. *J. Chem. Soc.* **1923,** *123*, 44.
(71) Gough, G.A.C.; Hunter, H.; Kenyon, J. *J. Chem. Soc.* **1926,** *128*, 2052.
(72) Kenyon, J. *Bull. Soc. Chim. Fr.* **1951,** C64.
(73) Walden, P. *Ber.* **1896,** *29*, 113.
(74) Polanyi, M.; Szabo, A.L. *Trans. Faraday Soc.* **1934,** *30*, 508.
(75) Holmberg, B. *Ber.* **1912,** *45*, 2997.
(76) Fischer, E. *Ann.* **1912,** *394*, 360.
(77) Hückel, W.; Frank, E. *Ann.* **1930,** *477*, 137.
(78) Verkade, P.E.; Coops, J, Jr.; Verkade-Sandbergen, A.; Maan, C.J. *Ann.* **1930,** *477*, 297.
(79) Kenyon, J.; Phillips, H.; Turley, H.G. *J. Chem. Soc.* **1925,** *127*, 399.
(80) Phillips, H. *J. Chem. Soc.* **1925,** *127*, 2552.
(81) Lapworth, A. *J. Chem. Soc.* **1904,** *85*, 30.
(82) Hammett, ref. 66, p. 96.
(83) Weissberger,A.; Dörkin, A.; Schwarze, W. *Ber.* **1931,** *64*, 1200.
(84) Weissberger, A.; Dym, E. *Ann.* **1933,** *502*, 74.
(85) Ingold, C.K.; Wilson, C.L. *J. Chem. Soc.* **1934,** *773*.
(86) Bartlett, P.D.; Stauffer, C.H. *J. Am. Chem. Soc.* **1935,** *37*, 2580.
(87) Ramberg, L.; Melander, A. *Arkiv. Chem. Mineral. Geol.* **1934 B,** *11*, No. 31. (b) Ramberg, L.; Hedlund, I. *ibid.* No. 41.
(88) (a) Hughes, E.D.; Ingold, C.K.; *J. Chem. Soc.* **1937,** *139*, 1196 and subsequent papers. (b) *ibid.* **1940,** *142*, 1010. (c) *Trans. Faraday. Soc.* **1938,** *34*, 202.
(89) Ingold, C.K. Memoir of Edward David Hughes, *Biogr. Mem. Fellows Roy. Soc.* **1964,** *10*, 147.

(90) Shannon, C.E.; Weaver, W. *The Mathematical Theory of Communication*, University of Illinois Press, Urbana, 1949, p. 100 ff.
(91) Eddington, A. as quoted in ref. 90.
(92) Weyl, H. *Symmetry*, Princeton University Press, Princeton, NJ, 1952. (a) p. 3. (b) p. 3–6. (c) p. 16. (d) p. 25–26. (e) p. 5. The stanza quoted therein is "Envoi", *from The Contemplative Quarry*, by Anna Wickham, Harcourt, Brace and Co., 1921. (f) See also refs. 93–95.
(93) For applications to chemistry of some of the themes Weyl treats in refs. 92a–d see *inter alia:* (a) Hargittai, I.; Hargittai, M. *Symmetry Through the Eyes of a Chemist*, 2nd ed. Plenum Press, New York, 1995. (b) *Reflections on Symmetry: in Chemistry and Elsewhere*, Heilbronner, E.; Dunitz, J.D. illustrations by Ruth Pfalzberger, Verlag Helvetica Chimica Acta, Basel; VCH Publishers,New York, 1993.
(94) Berlyne, D.E. *Aesthetics and Psychobiology*, Appleton-Century-Crofts, New York, 1971.
(95) Gombrich, E.H. (a) *The Sense of Order: A Study in the Psychology of Decorative Art*, Cornell University Press, Ithaca, 1979. (b) *The Image and the Eye: Further Studies in the Psychology of Pictorial Representation*, Phaidon, Oxford, 1982.
(96) Plato, *Symposium or Supper*, translated with an Introduction by Leslie, S. Fortune Press, London, 1933, p. 51 ff.
(97) Yeats, W.B. *Collected Poems of W.B. Yeats*, Finneran, R.J. ed. Macmillan, New York, NY, 1983.
(98) Plato, *Timaeus*, English and Greek. Archer-Hind, R.D., ed. Arno Press, New York, 1973.
(99) Kepler, J. *Mysterium Cosmographicum*, 1923 German translation (*Das Weltgeheimnis*) of the 1595 version. Translation and introduction by Caspar, M. Dr. Benno Filser Verlag Augsburg, 1923.
(100) A brief but pungent appreciation of Kepler's personality and intellectual qualities can be found in a notice by Newman (ref. 101a). A more extensive survey is given by Lodge (ref. 102).
(101) Newman, J.R. *The World of Mathematics*, Simon and Schuster, New York, 1956, Vol. I. (a) p.218. (b) p. 221.
(102) Lodge, Sir Oliver, as quoted in ref. 101.
(103) Cohen, I.B. in *Johannes Kepler. Mysterium Cosmographicum. The Secret of the Universe*, Duncan, A.M. (translation), Aiton, E.J (introduction and commentary), Cohen, I.B. (preface), Abaris Books, New York, 1981. (a) preface. (b) p. 7. (c) p. 30. (d) Kepler, J. *Le Secret du Monde*, Translation of *Mysterium Cosmographicum* into French by Segonds, A. Société d'Éditions "Les Belles Lettres," Paris, 1984.
(104) (a) "Bode's law," which astronomers used later to characterize the distance ratios, is described (ref. 101b) as follows:
» *Write down the series 0, 3, 6, 12, 24, 48 &c.; add 4 to each and divide by ten. The resulting series gives the approximate mean distances of the planets from the sun in astronomical units. For the six Copernican planets, the calculated (and actual) mean distances are Mercury, 0.4 (0.39); Venus, 0.7 (0.72); Earth, 1.0 (1.00); Mars,1.6 (1.52); Jupiter, 5.2 (5.20); and Saturn, 10.0 (9.53). This is the law discovered by the German astronomer Johann Elert Bode (1747-1826) in 1772. Its failure in the case of Neptune and Pluto, planets found after Bode's time, has led most astronomers to conclude that the 'law' is a purely empirical relationship, more in the realm of coincidence than an actual physical law. Besides not being a law, it was not in fact discovered by Bode but by the German mathematician J.D. Titius (1729–1796).* «

(105) Hawking, S. *A Brief History of Time: From the Big Bang to Black Holes*, Bantam Books, New York, NY, 1988, p. 175.
(106) (a) Ref. 99, pp. I–XXXI. (b) ref. 101, Vol. 2, p.1316.
(107) Pierre Simon, Marquis de Laplace, renowned French astronomer and mathematician, 1749–1827. According to Newman,[106b] Laplace was
»*second only to Newton as a mathematical astronomer and mathematician; as a person, his qualities were mixed ... Above all, he was a virtuoso in the art of rapid adaptation to a changing social and political environment. He is condemned for this – probably because it is such a common frailty that mere mention of it makes everyone uncomfortable.*«
(108) (a) Bally, T.; Masamune, S. *Tetrahedron*, **1980,** *36*, 343. and references cited therein. (b) Whitman, D.W.; Carpenter, B.K. *J. Am. Chem. Soc.* **1980,** *102*, 4272. (c) Schröder, G. *Cyclooctatetraene*, Verlag Chemie, Weinheim, 1965 and references cited therein. (d) Traetteberg, M. *Acta Chem. Scand.* **1966,** *20*, 1724. (e) Buma, W.J.; van der Waals, J.H.; van Hemert, M.C. **1989,** *111*, 86.
(109) (a) Samuels, C.A.; Butterworth, G.; Roberts, T.; Graupner, L.; Hole, G. *Perception*, **1994,** *23*, 823. (b) Markow, T.A.; Bustoz, D.; Pitnick, S. *Animal Behavior*, **1996,** *52*, 759.
(110) (a) Herbert, A.M.; Humphrey, G.K. *Perception*, **1996,** *25*, 463. (b) References cited in (a).
(111) Enquist, M.; Arak, A. *Nature*. (a) **1994,** *372*, 169. (b) **1995,** *374*, 313.
(112) Cook, N.D. *Nature*, **1995,** *374*, 313.
(113) (a) Johnstone, R.A. *Nature*. (a) **1994,** *372*, 172. (b) **1995,** *374*, 313.
(114) Ref. 94, p. 14.
(115) Ref. 94, p. 16.
(116) Ref. 94, pp. 16-17.
(117) Ref. 94, p. 1.
(118) Fiske, P.; Amundsen, T. *Animal Behavior*, **1997,** *54*, 81.
(119) Swaddle, J.P.; Cuthill, I.C. *Proc. Roy. Soc. B Biol. Sci.* **1995,** *261*, 111.
(120) Chen, A.C.; German, C.; Zaidel, D.W. *Neuropsychologia*, **1997,** *4*, 471.
(121) Kowner, R. *J. Exp. Psychol. Hum. Percept. Perform.* **1996,** *22*, 662.
(122) (a) Mach, E. *The Analysis of Sensations and the Relation of the Physical to the Psychical* (revised and supplemented from the fifth German edition) Dover, New York, 1959. (First edition 1886). (b) For a brief appreciation of Mach's contributions to physics, the philosophy of science, and physiology, see Szasz, T.S., introduction to ref. 122a, pp. v–xxxi. (c) See also Janik, A.; Toulmin, S. *Wittgenstein's Vienna*, Simon and Schuster, New York, 1973, p. 133 and elsewhere. These authors summarize Mach's impact upon world culture in the words:
»*From poetry to philosophy of law, from physics to social theory, Mach's influence was all-pervasive in Austria and elsewhere. Not the least of those who came under Mach's spell was the young physicist Albert Einstein, who acknowledged Mach's 'profound influence' upon him in his youth. It has further been brought to light that Einstein's early career was modeled on Mach's view of the nature of the scientific enterprise. After meeting Mach, a dazzled William James could only refer to him as a 'pure intellectual genius,' who read and discussed absolutely everything.*«
(d) See also Blackmore, J.T. *Ernst Mach: His Life, Work, and Influence*, University of California Press, Berkeley, CA, 1972, pp. 29–30, 48ff.
(123) Julesz, B. *Foundations of Cyclopean Perception*, University of Chicago Press, Chicago, 1971.
(124) Various authors cited in ref. 110.
(125) Tyler, C.W. *Nature*, **1998,** *392*, 877.

Chapter 6

Epilogue

The *Harvard Case Studies in the History of Science*, which we encountered in Chapter 1, used techniques similar to those we have used here, in that they analyzed actual investigations in a detailed way. However, their aims were to teach the lay public enough about how science is done in order to facilitate the making of reasoned judgments about public policy. In this book, however, we have addressed professional chemists. Why have we busied ourselves here with these stories of long-ago struggles? Why don't we just accept those things that are certain and move ahead? After all, do our teachers not tell us that the old, packed-down, forgotten chemical literature is like a forest floor, on which the dried leaves and vine tendrils and rotted stumps of bygone years form a richly nutrient matrix which sustains new life? Why dig up all those tangled layers? There are several reasons we have not yet explicitly pointed out.

First is the matter of *connoisseurship*, in the sense of its Latin root *cognoscere*, to know.[1a] This is not to be taken as cultivation of the antique or the quaint. Rather, as professionals in this discipline, we are persons who are supposed to be "competent to act as a critical judge of an art or, in a matter of taste." This competence is not solely to help in the actual prosecution of our own scientific reasearch, but also for the judgment of the research of others. It is the basis of the whole system of peer review upon which depends the disbursement of government and private funds for the support of research. Unless we appreciate, at least to some degree, how we got to this point, it will be difficult to discern the relationship of the parts of the discipline to the whole. Unless we understand how subdisciplines that now seem superficially disparate grew out of common roots (as, for example, we have seen to be the case with natural products and reaction mechanisms), we will not be able to pass on to our successors the means to evaluate work outside our own narrow specialties.

A second reason comes from an appreciation of our *inheritance*, which in the admirable dictionary definition[1b] is seen as "a permanent or valuable possession or blessing, esp. one received by gift or without purchase; a benefaction." The act of immersing ourselves vicariously in the milieu of the time of a scientific advance, while thinking within the limitations of what was known at that point in the history of the subject, leads us not only to appreciate the magnitude of our predecessors' achievements. It also sharpens our self-awareness and teaches us to be alert for what it is that we today surmise to be true or accept as proven. We saw, for example, in the study of the Wagner rearrangements, that the argument at one time was cast as a bipolar choice: either the α-elimination mechanism or the tricyclene mechanism. In those terms, once the α-elimination mechanism had been shown to be incorrect, it then followed that the tricyclene mechanism must be correct. We recognize today that the fallacy in this procedure was the restriction of the issue to only two possible answers. How many mechanistic arguments today will be (temporarily) "decided" by a similar logical error? A knowledge of the past not only can shield us from a too-ready acceptance of interpretations, but also can inspire us to create alternatives.

Related to these is the value of developing a sensitivity to just what kinds of information we find most persuasive. We saw, for example, the powerful attraction of symmetry in conditioning our receptivity to observations and interpretations. Although we have not been able to discern here the deep psychological roots of that issue, I think that by exploring it, we may have encouraged some introspection and curiosity about how our minds work. This self-knowledge may deter us in the future from blunders caused by seductive biases inherent in the human psyche itself.

Reference

(1) Webster's Collegiate Dictionary, 5th ed. G.C. Merriam, Springfield, MA, 1948: (a) p. 215. (b) p. 517.

Author Index

References at the end of each chapter are listed below in the form: chapter number (page of citation)[reference number].

Subject Index